Feature Papers of Forecasting

Feature Papers of Forecasting

Editor

Sonia Leva

MDPI • Basel • Beijing • Wuhan • Barcelona • Belgrade • Manchester • Tokyo • Cluj • Tianjin

Editor
Sonia Leva
Department of Energy, Politecnico Di Milano,
Via Lambruschini 4
Italy

Editorial Office
MDPI
St. Alban-Anlage 66
4052 Basel, Switzerland

This is a reprint of articles from the Special Issue published online in the open access journal *Forecasting* (ISSN 2571-9394) (available at: https://www.mdpi.com/journal/forecasting/special_issues/FP_Forecast).

For citation purposes, cite each article independently as indicated on the article page online and as indicated below:

LastName, A.A.; LastName, B.B.; LastName, C.C. Article Title. *Journal Name* **Year**, *Volume Number*, Page Range.

ISBN 978-3-0365-1030-9 (Hbk)
ISBN 978-3-0365-1031-6 (PDF)

Contents

About the Editor

Sonia Leva is Full Professor in "Elettrotecnica" (Electrical Engineering-Circuit Theory) in Politecnico di Milano (Italy). She was born in 1970 in Tradate (Italy). She received the M.Sc. degree in 1997, and the Ph.D. degree in 2001, both in Electrical Engineering, from the Faculty of Engineering, Politecnico di Milano, Italy. In 1998, she registered as a professional Engineer in Italy. From 1999 to 2010, she was Research Associate of Electrical Engineering at the Department of Electrical Engineering, Politecnico di Milano, Italy. Since July 2010, Sonia Leva is qualified as Associate Professor, starting her professorship on December 16, 2010. In February 2014, Sonia Leva qualified as a Full Professor in "Elettrotecnica" (Electrical Engineering-Circuit Theory), starting her professor activity on January 06, 2016.

She has been an IEEE member since 2000 and a senior member since 2013. She served as a Chairperson of sessions in an international conference organized by Institute of Electrical and Electronic Engineers. She is the author of about 100 papers mainly published on international and national journal or conference proceedings. She served as Editor-in-Chief for Forecasting from 2019.

Sonia Leva is member of the Italian Standard Authority (CEI) Technical Committee CT 82 "Sistemi di conversione fotovoltaica dell'energia solare (Photovoltaic Systems)" since 2008. She collaborated to write the second edition of the technical guide.

 forecasting

MDPI

Editorial

Editorial for Special Issue: "Feature Papers of Forecasting"

Sonia Leva

Department of Energy, Politecnico di Milano, 20156 Milano, Italy; sonia.leva@polimi.it

Nowadays, forecasting applications are receiving unprecedent attention thanks to their capability to improve the decision-making processes by providing useful indications. A large number of forecasting approaches related to different forecasting horizons and to the specific problem that have to be predicted have been proposed in recent scientific literature, from physical models to data-driven statistic and machine learning approaches. Hybrid approaches combining two or more of the previously-mentioned methods, have been also investigated. In general, two methods based on AI-Based Techniques.

In this Special Issue, the most recent and high-quality researches about forecasting are collected. A total of nine papers have been selected to represent a wide range of applications, from weather and environmental predictions to economic and management forecasts. Finally some application related to forecasting the different phase of COVID in Spain and the photovoltaic power production have been presented.

Pedrigão et al. [1] compare the Direct Normal Irradiance (DNI) predictions over one year (from 1 August 2018 to 31 July 2019) from Integrated Forecasting System of European Centre for Medium-Range Weather Forecasts (IFS/ECMWF) against the corresponding observed values in south of Portugal, in Évora station, for different time steps (hourly and daily basis) and for different forecast horizons (up to four days ahead). The comparison highlights similar magnitude and trend between forecast and observed data, with an over-estimation of the predicted DNI by IFS/ECMWF and a general error increase with forecast lead time. In addition, a methodology based on DNI attenuation index (DAI) is proposed to estimate the transparency of the atmosphere and a post-processing methodology is adopted to reduce the bias in IFS/ECMWF model. The first methodology revealed the tendency of IFS/ECMWF approach to underestimate the effects of clouds on DNI, while the bias correction post-processing allowed a large improvement in the DNI forecast, with a 30% decrease for all error metrics.

Maroccu et al. [2] propose a nowcasting method for precipitation intensity predictions. The method is based on a generative neural network, which is trained with a specific loss function and presents a PredNet architecture, successfully applied in other fields such as computer vision and natural language processing. Its forecast performance is compared against those of state-of-art optical flow procedures in a real case study, a public domain dataset of radar images from Japan covering a time span of five years. The results demonstrate the neural network to be by far the most effective forecast model between all those proposed.

Gunter et al. [3] investigate the accuracy of individual and combined statistical methods in forecasting the tourism demand for the European Union. This research, contracted by the European Commission, required the analysis of models with low degree of complexity, easily rebuildable for a practical application. The accuracy assessment was performed for eight different periods spanning two years each, in order to grant stable results inside a changing macroeconomic context. The results demonstrate that the combination between Autoregressive Integrated Moving Average (ARIMA) models, REGARIMA models and Error Trend Seasonal (ETS) models performs better than single models, and that the combination based on Bates–Granger weights (VAriance-COvariance methods, VACO) is better performing than the one based on uniform weights.

Ghimire et al. [4] analyze the accuracy of low-complexity data-driven persistence-based approaches for streamflow forecasting in Nepal, in the Himalayan region, proposed

Citation: Leva, S. Editorial for Special Issue: "Feature Papers of Forecasting". *Forecasting* **2021**, *3*, 135–137. https://doi.org/10.3390/forecast3010009

Received: 18 February 2021
Accepted: 20 February 2021
Published: 21 February 2021

as benchmark for the real-time streamflow forecasting system. In detail, a simple persistence approach, a streamflow climatology approach and an anomaly persistence approach are discussed. In general, the forecast skill of persistence-based methods presents a strong spatio-temporal dependence: it is higher in rivers with constant baseflow respect to intermittent ones, with moderate flows respect to extreme ones and with larger river basins respect to smaller ones. Finally, the study demonstrates that the proposed persistence-based forecast approaches are difficult to outperform even with complex mechanistic hydrologic models.

Rezazadeh [5] proposes a Machine-Learning (ML) workflow implemented on the cloud-based computing platform Microsoft Azure Machine-Learning Service (Azure ML) with the aim of predicting the possibilities of winning sales opportunities in Business to Business (B2B) sales, a task nowadays mostly relying on human evaluations. In order to investigate the effectiveness of the discussed approach, it was applied to a multi-business consulting firm: the ML workflow performance was compared with user-entered predictions made by salespersons. The results demonstrate the decision-making based on the ML predictions to be more accurate than the subjective human predictions.

Comi et al. [6] investigate the variability of bus travel time by means of a time-series-based approach applied on data from Automated Vehicle Monitoring (AVM) of bus lines sharing the road lanes with traffic (private vehicles) in Rome (Italy) and Lviv (Ukraine). The analysis of the results point out the efficiency of the proposed forecast approach, highlighting also the significant effects of time of the day and the day of the week on travel time variability. Moreover, also the key structural differences between Rome and Lviv are considered, showing the need to account them when developing a forecast model.

Hogan et al. [7] describe a diminishing learning rate model, namely Boone's learning curve, with the aim to improve the end-items cost evaluation and to propose an alternative to popular but outdated learning curve models adopted by U.S. Department of Defense (DoD) in costs estimation. The research demonstrates that Boone's learning curve significantly reduces error in modeling observed learning curves using production data from 169 DoD end-items.

Mora et al. [8] develop and calibrate a semi-empirical model based on logistic maps with the aim to forecast the different phases of the COVID-19 epidemic in Spain. In detail, the model predicts the number of infected, hospitalized, patients needing an Intensive Care Unit (ICU) and deaths in four different epidemic phases, namely: non-controlled evolution, total lock-down, partial easing of the lock-down and phased lock-down easing. A reliable forecast of COVID-19 development for both countries or smaller regions allows an optimization of sanitary resources and permits the reduction of the economic and social impact of Non Pharmaceutical Interventions (NPIs), such as lock-downs. The proposed model was capable to provide reasonably accurate results for the different phases of the epidemic.

Pan et al. [9] propose a data-driven Photovoltaic (PV) output power estimation approach using only net load data, temperature data and solar irradiation data and involving a decomposition method of the Behind-The-Meter (BTM) PV output power curve. In order to illustrate the effectiveness of the described approach, the PV output decomposition was simulated and tested on a total of 300 real customers' datasets from Ausgrid.

Conflicts of Interest: The authors declare no conflict of interest.

References

1. Perdigão, J.; Canhoto, P.; Salgado, R.; Costa, M.J. Assessment of Direct Normal Irradiance Forecasts Based on IFS/ECMWF Data and Observations in the South of Portugal. *Forecasting* **2020**, *2*, 130–150. [CrossRef]
2. Marrocu, M.; Massidda, L. Performance Comparison between Deep Learning and Optical Flow-Based Techniques for Nowcast Precipitation from Radar Images. *Forecasting* **2020**, *2*, 194–210. [CrossRef]
3. Gunter, U.; Önder, I.; Smeral, E. Are Combined Tourism Forecasts Better at Minimizing Forecasting Errors? *Forecasting* **2020**, *2*, 211–229. [CrossRef]

4. Ghimire, G.R.; Sharma, S.; Panthi, J.; Talchabhadel, R.; Parajuli, B.; Dahal, P.; Baniya, R. Benchmarking Real-Time Streamflow Forecast Skill in the Himalayan Region. *Forecasting* **2020**, *2*, 230–247. [CrossRef]
5. Rezazadeh, A. A Generalized Flow for B2B Sales Predictive Modeling: An Azure Machine-Learning Approach. *Forecasting* **2020**, *2*, 267–283. [CrossRef]
6. Comi, A.; Polimeni, A. Bus Travel Time: Experimental Evidence and Forecasting. *Forecasting* **2020**, *2*, 309–322. [CrossRef]
7. Hogan, D.; Elshaw, J.; Koschnick, C.; Ritschel, J.; Badiru, A.; Valentine, S. Cost Estimating Using a New Learning Curve Theory for Non-Constant Production Rates. *Forecasting* **2020**, *2*, 429–451. [CrossRef]
8. Mora, J.C.; Pérez, S.; Dvorzhak, A. Application of a Semi-Empirical Dynamic Model to Forecast the Propagation of the COVID-19 Epidemics in Spain. *Forecasting* **2020**, *2*, 452–469. [CrossRef]
9. Pan, K.; Xie, C.; Lai, C.S.; Wang, D.; Lai, L.L. Photovoltaic Output Power Estimation and Baseline Prediction Approach for a Residential Distribution Network with Behind-the-Meter Systems. *Forecasting* **2020**, *2*, 470–487. [CrossRef]

Article

Assessment of Direct Normal Irradiance Forecasts Based on IFS/ECMWF Data and Observations in the South of Portugal

João Perdigão [1,*], Paulo Canhoto [1,2], Rui Salgado [1,2] and Maria João Costa [1,2]

1 Instituto de Ciências da Terra, Universidade de Évora, Rua Romão Ramalho 59, 7000-671 Évora, Portugal; canhoto@uevora.pt (P.C.); rsal@uevora.pt (R.S.); mjcosta@uevora.pt (M.J.C.)

2 Departamento de Física, Escola de Ciências e Tecnologia, Universidade de Évora, Rua Romão Ramalho 59, 7000-671 Évora, Portugal

* Correspondence: perdi.j@gmail.com

Received: 19 April 2020; Accepted: 14 May 2020; Published: 16 May 2020

Abstract: Direct Normal Irradiance (DNI) predictions obtained from the Integrated Forecasting System of the European Centre for Medium-Range Weather Forecast (IFS/ECMWF) were compared against ground-based observational data for one location at the south of Portugal (Évora). Hourly and daily DNI values were analyzed for different temporal forecast horizons (1 to 3 days ahead) and results show that the IFS/ECMWF slightly overestimates DNI for the period of analysis (1 August 2018 until 31 July 2019) with a fairly good agreement between model and observations. Hourly basis evaluation shows relatively high errors, independently of the forecast day. Root mean square error increases as the forecast time increases with a relative error of ~45% between the first and the last forecast. Similar patterns are observed in the daily analysis with comparable magnitude errors. The correlation coefficients between forecast and observed data are above 0.7 for both hourly and daily data. A methodology based on a new DNI attenuation Index (DAI) was developed to estimate cloud fraction from hourly values integrated over a day and, with that, to correlate the accuracy of the forecast with sky conditions. This correlation with DAI reveals that in IFS/ECMWF model, the atmosphere as being more transparent than reality since cloud cover is underestimated in the majority of the months of the year, taking the ground-based measurements as a reference. The use of the DAI estimator confirms that the errors in IFS/ECMWF are larger under cloudy skies than under clear sky. The development and application of a post-processing methodology improves the DNI predictions from the IFS/ECMWF outputs, with a decrease of error of the order of ~30%, when compared with raw data.

Keywords: Direct Normal Irradiance (DNI); IFS/ECMWF; forecast; evaluation; DNI attenuation Index (DAI); bias correction

---

1. Introduction

Solar energy is becoming a crucial renewable resource in modern societies, contributing to the sustainability of the planet with the mitigation of greenhouse gas emissions by reducing the consumption of coal or fuel oil for electricity production; however, the availability of solar resources over time at a given region of interest determines the cost/benefit of solar power plants implementation. Since the temporal series of solar radiation measurements are spatially limited, and thus scarce and sometimes inexistent, the prediction and validation of solar resource is a key factor for such enterprises.

Several researchers have estimated the potential of renewable energies like wind or solar radiation for electricity or thermal energy production around the world; for example, in Europe and Africa [1], in Chile [2], in Iberian Peninsula [3], in United Kingdom [4], and Spain [5]. Solar power is a very promising energy source in the Iberian Peninsula (IP) and strong growth is expected in this area. In the IP there are multiple options for using renewable energy (solar, wind, and hydro) to generate electricity; however, the solar resource is high throughout the year [6–8].

Concerning solar energy, there are two main ways of converting solar energy into electricity: Photovoltaic (PV) and Concentrating Solar Power (CSP). The PV panels convert either direct and diffuse solar irradiance, while the CSP technology only concentrates the Direct Normal Irradiance (DNI). The focus of this work is on the prediction of DNI, because of its use in CSP plant management. The forecast of global solar radiation (direct + diffuse) for the same region was addressed, for example, in [6] and [9].

There are several approaches to predict solar irradiance such as Numerical Weather Prediction (NWP), Cloud Motion Vector (CMV), statistical time series analysis, and other methods [7,10,11]. In the last years, one of the major research challenges for the use of NWP in solar energy applications is the DNI forecast, aiming at the development and increase of CSP installed capacity and operation management. CSP requires the knowledge of DNI for specific sites [12,13], and one of the difficulties is the need to forecast the DNI with several days ahead to increase efficiency and minimize the operational costs of the power plants [14,15]. For instance, Casado-Rubio et al. [16] proposed a simple methodology to obtained DNI forecast, based on Weather Research and Forecasting model (WRF model) and a radiative transfer simulation (for 1-day forecast) and found that this procedure can be used as a diagnostic tool for operational power plants.

Until recently, DNI measurements were not available in many places or with long series, and this variable was not a direct output of NWP models. Currently, the Integrated Forecast System of the European Centre for Medium-Range Weather Forecasts (IFS/ECMWF) provides the direct normal irradiance as an output; however, the use of NWP models in the DNI forecast is still not perfect and requires Multiple Output Statistic (MOS) methodologies [13]. Lopes et al. [17] used the IFS global model of ECMWF to assess DNI for short-term (24 h) in the south of Portugal and found relative differences in the range ~7% to ~12% on an annual basis between predictions and observations at ground-based stations. Lara-Fanego et al. [18] found a relative root mean square error of 60% for hourly DNI forecasts in Spain for all sky conditions, using the Advanced Research Weather Research and Forecasting model (WRF). Troccoli and Morcrette [19] analyzed the direct solar radiation data using two different radiation schemes of the IFS/ECMWF for four ground-based measuring stations in Australia and found mean absolute errors between 18% and 45% and correlation coefficients between 0.25 and 0.85. In that work, the usage of a post-processing bias correction improved results, resulting in mean absolute errors between 10% and 15% and correlation coefficients of about 0.9. Ruiz-Arias et al. [7] also found better results for DNI forecasts from the WRF model by using a post-processing algorithm. Law et al. [13] present a comprehensive review of DNI forecast obtained from several methods and some examples of DNI forecast accuracies are presented. According to Vick et al. [20], most studies on DNI models have assessed the annual and hourly mean bias and root mean square errors between measured and DNI models; however, according to the same authors, the accuracy of monthly and daily direct normal irradiation forecasts should also be assessed to detect gaps in DNI modeling that may be improved and correlated with sky conditions, time of the year, or location.

Since there are several on-going projects in Portugal to explore the solar resource, it is imperative to carry out studies that help understand the errors associated with direct normal irradiance predictions over several days ahead.

The Portuguese Institute for Sea and Atmosphere (IPMA), the meteorological Portuguese authority, uses the ECMWF global model predictions as the main forecast tool. Comparisons between operational global NWP models show that ECMWF over Europe is the best [21]. Good numerical predictions of the near-surface weather conditions presuppose a good representation of the surface radiative balance; however, a correct forecast of the global irradiance does not necessarily mean an accurate partition between direct and diffuse components, as this partition is not essential to solving the surface balance.

In response to growing demand from the solar energy market, the ECMWF has recently (2015) started to include DNI among the available predicted variables. These forecasts are likely to be used by solar plants in southern Portugal.

The main objective of this study was to assess the performance of the IFS/ECMWF global model (CY45R1 cycle—released at 5 June 2018) to predict DNI in the south of Portugal, by comparing its results with observational data of Évora station, on an hourly and daily basis and for various forecasting horizons (up to four days ahead). We present a method to predict sky conditions based on observational DNI data. A post-processing methodology was also tested to minimize the bias in the IFS/ECMWF model.

The paper is organized as follows: Section two describes the data and methods used in this study, the performance assessment of the ECMWF forecasts is presented and discussed in Section 3, and finally, conclusions are provided in Section 4.

2. Materials and Methods

2.1. DNI Observational Data

The measurements used in this study were obtained from the observatory of Atmospheric Sciences located at the University of Évora (38.57° N, 7.9° W, 293 m a.m.s.l.). DNI was measured using a first-class pyrheliometer (Kipp & Zonen, model CHP01) [22], following the World Meteorological Organization (WMO) [23] and the International Organization for Standardization (ISO), the 9060:1990 standard [24]. This model of pyrheliometer was designed to measure the solar irradiance with an opening half-angle of 2.5°.

A period of one year of DNI measurements was used in this study, from 1 August 2018 until 31 July 2019. The sensor output was sampled every 5 s and one-minute mean, minimum, maximum, and standard deviation values were recorded. Hourly values were then computed by averaging one-minute values when the number of records for that hour corresponds to at least fifty minutes. The data for solar zenith angles above 89° (twilight and nighttime) were not considered and thus removed from the analysis. The daily mean was computed using a similar methodology of that used by Troccoli and Morcrette [19], i.e., if one or more hourly values are not present on a given day, then that day is not used in the analysis.

All instruments of this measuring station were subject to maintenance and cleaning procedures following the recommendations of the World Meteorological Organization and data was subject to BSRN (Baseline Surface Radiation Network) quality filters [25] based on physically possible and extremely rare values.

In this work, the seasons were defined according to the WMO nomenclature i.e., winter (December–January–February: DJF), spring (March–April–May: MAM), summer (June–July–August: JJA), and autumn (September–October–November: SON).

2.2. DNI Forecast Data

Predicted DNI from IFS/ECMWF, from 1 August 2018 until 31 July 2019, was obtained with a resolution of 0.125 × 0.125 (lat × lon grid). The forecast data were provided with an hourly time step for the first three days and with a three-hour time step for the 4th day. In this work, forecasts were separated into four intervals: day_0 (1st day); day_1 (2nd day); day_2 (3rd day); day_3 (4th day). The predicted accumulated solar irradiation in hour time steps for the entire forecast horizon was converted into hourly mean irradiance values of DNI.

The shortwave radiation scheme of the IFS/ECMWF used in this study was the new radiation scheme implemented on 11 July 2017, called ecRad [26]. This scheme is faster than the previously McRad scheme [27] and can be executed more times during the forecast. This scheme computes the profiles of shortwave and longwave irradiances at half levels, and these are interpolated horizontally back onto the model grid using cubic interpolation [26]. The aerosol distribution was adapted from

Tegen et al. [28], using a climatology of six hydrophobic aerosol species as well as the newer climatology obtained from a reanalysis of the atmospheric composition produced by the Copernicus Atmosphere Monitoring Services (CAMS), with 11 hydrophilic and hydrophobic species [29].

According to Hogan and Bozzo [26], ecRad incorporates a method to represent longwave scattering of clouds, which leads to an improvement in forecast skills. The default ice optical properties were computed using the Fu scheme [30], but two additional schemes were available.

In the work by Hogan and Bozzo [26], it was possible to find the evolution of the ECMWF Radiation Scheme after 2000 and the options available. More details on the physical processes (and the options available) were reported in the IFS documentation on the ECMWF web page [31].

The nearest neighbor technique was used to select the forecast data for comparison with measurements. To assess forecasting accuracy, the observational data was compared with the forecasts for the nearest model grid point. Wild and Schmucki [32], made several statistical tests surrounding a grid point to analyze trends and the results showed that different grid points surrounding a given grid point (selected by a Lat/Lon value) do not differ significantly from each other in the majority of the cases.

2.3. *Statistical Indicators for Model Sssessment*

The quality of the DNI forecasts was evaluated against observational data using common statistical parameters as the Mean Bias Error (MBE), Mean Absolute Error (MAE), Root Mean Square Error (RMSE), and Correlation Coefficient (r). In this work, errors were calculated based on hourly, daily, and monthly mean values. Similar to the analysis presented by Nonnenmacher et al. [14] and Perez et al. [33], night-time values (zero solar irradiance) were excluded from the model assessment. The ratio (RSR) between the root mean square error and the observations standard deviation (σ_{obs}) was also determined.

These statistical parameters are defined as follows:

$$MBE = \frac{1}{N}\sum_{i=1}^{N}(m_i - o_i) \tag{1}$$

$$MAE = \frac{1}{N}\sum_{i=1}^{N}|m_i - o_i| \tag{2}$$

$$RMSE = \left[\frac{1}{N}\sum_{i=1}^{N}(m_i - o_i)^2\right]^{\frac{1}{2}} \tag{3}$$

$$r = \frac{\sum_{i=1}^{N}(o_i - \overline{o})(m_i - \overline{m})}{\left[\sum_{i=1}^{N}(o_i - \overline{o})^2 \sum_{i=1}^{N}(m_i - \overline{m})^2\right]^{\frac{1}{2}}} \tag{4}$$

$$RSR = \frac{RMSE}{\sigma_{obs}} = \frac{\left[\frac{1}{N}\sum_{i=1}^{N}(m_i - o_i)^2\right]^{\frac{1}{2}}}{\left[\sum_{i=1}^{n}(o_i - \overline{o})^2\right]^{\frac{1}{2}}} \tag{5}$$

where N is the number of data points and m and o are the forecast and observed values, respectively. The MBE represents a systematic error between predicted and observational values, and the RMSE quantifies the spread in the distribution of errors. The MBE provides information on the underestimation (negative values) or overestimation (positive values) of forecasts using the measured values as reference. On the other hand, the RMSE is very sensitive to high magnitudes errors due to the higher statistical weight of large errors. The MAE represents the average magnitude of errors in a set of forecasts without considering their direction (bias) and gives the same weight to all errors (see as an example, Chai and Draxler [34]), i.e., it is less sensitive to large deviations. RMSE is one of the most relevant statistical parameters for solar power plant analysis (e.g., Landelius et al. [35]). For all statistical parameters, the best results are obtained when values are equal or near zero, except for the correlation coefficient when values closer to one correspond to better performances. According to Moriasi et al. [36], RSR incorporates the benefits of error index statistics and includes a scaling/normalization factor, so that the resulting statistic and reported values can apply to various constituents. The RSR parameter varies between zero and a positive value with the values close to zero representing a better forecast simulation [36]. The same thresholds performance ratings as shown in Table 4 of the article of Moriasi et al. [36] are used here, i.e., values of RSR < 0.5 indicate the optimal performance rating while RSR > 0.7 represents unsatisfactory model performance rating.

2.4. Cloud Area Fraction and DNI Attenuation Index (DAI)

In most of the solar radiation studies, an index known as clear sky index or the clearness index is used to quantify the bulk atmosphere transmittance (see Iqbal [37], Lopes et al. [17], among others). This clearness index is defined as the ratio of global horizontal irradiance to extraterrestrial horizontal irradiance.

In this section, a new methodology is proposed to estimate the clearness of the atmosphere (termed DNI attenuation index-DAI) based exclusively on the observed direct normal irradiance. DNI varies during the day due to Sun–Earth geometry and atmospheric constituents, though the main factor of DNI variation is the cloud coverage, which can drastically reduce this component of solar radiation when the direct beam from the sun is intercepted, sometimes reaching a value of zero depending on the type of clouds. The DAI is an indicator of the cloud attenuation of DNI.

This method was based on the integration of the measured hourly mean values of DNI (see Figure 1a) obtained every day, for a given month, and it constitutes a measure of sky conditions for a particular month. This has the advantage of not relying on reanalysis, satellite data, or other products that could also be a source of bias.

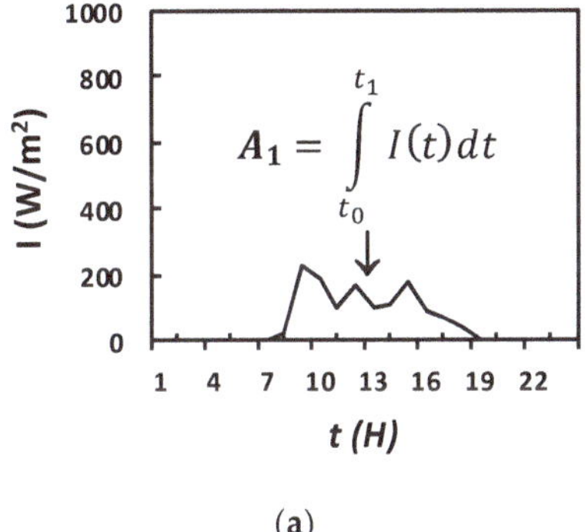

(a)

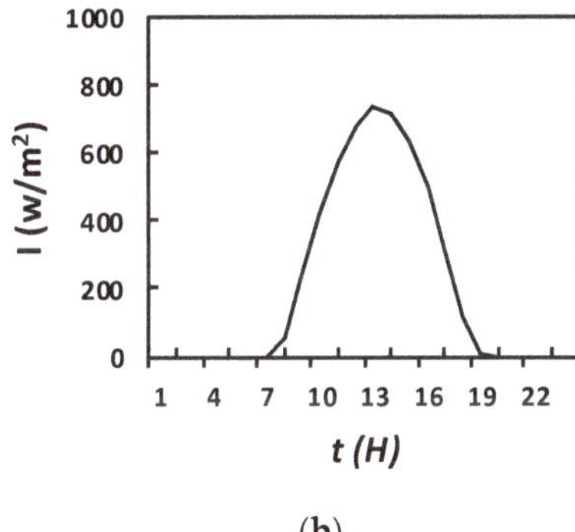

(b)

Figure 1. Curves of hourly mean Direct Normal Irradiance (DNI) projected on the horizontal plane for two different days: (**a**) partially cloudy sky and (**b**) clear sky day. A is the area under the curve corresponding to the measured DNI (energy per unit area) and is obtained by numerical integration using the trapezoid rule.

In this way, a dimensionless quantity (in percentage) called DNI Attenuation Index (DAI) is defined as

$$DAI_i = \left(1 - \frac{A_i}{NF}\right) \times 100\% \tag{6}$$

where

$$A_i = \int_{t0}^{t1} I(t)dt \sim \frac{\Delta t}{2} \sum_{k=1}^{24} (I_{k-1} + I_k) \tag{7}$$

with Δt = 3600 s because a time step of one hour is used and NF is the normalization factor calculated as

$$NF = \max_{1 \leq i \leq n}(A_i) \tag{8}$$

in which i is the number of the day of the month.

To obtain the DAI it is assumed that the maximum value of daily energy per unit area (integral) in a given month is interpreted as a clear sky day in that month and the DAI will take the value of zero for that particular day. Different normalization factors will be expected for different months, with higher values during the summer. Although DAI does not allow us to effectively distinguish the contribution of aerosols or cloud cover to DNI variations, it provides a clear idea of the transparence of the atmosphere for a specific day and it hints at the identification of a clear day (or clearness of atmosphere) from an overcast day or extreme aerosol event. The DAI varies between zero (clear sky day) and one (overcast sky).

The relation between DAI and cloud fraction (in oktas) was established through three classes of days [23]: class I – clear sky day (0–2 oktas; DAI< 31.25%); class II – partially cloudy skies (3–5 oktas; 31.25% ≤DAI< 68.75%) and class III as cloudy skies (6–8 oktas; 68.75% ≤ DAI ≤ 100%), in the same way as presented in Table 1 of the article of Jafariserajehlou et al. [38].

Table 1. Statistical indicators of comparison between observed and predicted hourly mean Direct Normal Irradiance (DNI) for the entire period (1 August 2018–31 July 2019). Bold values mean the best score.

Day	MBE (W/m^2)	MAE (W/m^2)	RMSE (W/m^2)	r
0	13.54	**136.80**	**195.41**	**0.84**
1	15.03	146.35	210.60	0.81
2	17.273	154.97	224.02	0.78
3	**1.048**	197.88	267.25	0.70

The total cloud area fraction obtained from the Clouds and the Earth's Radiant Energy System (CERES) radiometer, combined with the Moderate Resolution Imaging Spectroradiometer (MODIS), both onboard the Terra and Aqua satellites, was also considered in this work for assessment of DAI estimates. The CERES–MODIS cloud mask data were obtained monthly (CERES_SYN1deg_Ed4.1) for the period available for this study (August 2017 until May 2019) and from CERES portal (see ceres.larc.nasa.gov). The cloud area fraction consists of the percentage of cloudy pixels identified in areas of 1° × 1° [39].

2.5. Post Processing Correction

A linear least square statistical method for bias correction to correct daily direct solar radiation values obtained from IFS/ECMWF was tested. This method is the simplest post-processing technique and has been applied in several studies over the past years (see for example Polo et al. [40]). Mejia et al. [41] found that MOS linear fit procedure outperformed the quantile–quantile mapping (Q–Q).

The linear regression parameters were computed for each month using forecast and observed daily values for the period of 1 August 2017 until 31 July 2018. The correction parameters are obtained, for each month, using the linear equation

$$\begin{gathered} y^i_{model,j} = m_j x^i_{obs,j} + b_j \\ i = 1, \ldots 28/30/31; j = 1, \ldots 12 \end{gathered} \tag{9}$$

where x_{obs}, m_j and b_j are, respectively, the observed DNI values, the slope of the fitted line, and the intercept.

The regression parameters were used to correct the IFS/ECMWF forecasts for the following year—the period of analysis (01/08/2018 until 31/07/2019)—using the following equation [42],

$$y^i{}_{BC_{model},j} = y^i{}_{model,j} - \left[\left(m_j - 1\right)x^i{}_{obs,j} + b_j\right] \quad i = 1, \ldots 28/30/31; j = 1, \ldots 12 \tag{10}$$

3. Results and Discussion

3.1. Assessment of Hourly and Daily DNI Forecasts

As an example, Figure 2 shows the time series of predicted hourly mean DNI during four consecutive days and the corresponding observed values for two selected cases, one in JJA, other in SON: forecasts issued on 1 March 2017 00:00 and on 27 November 2017 00:00.

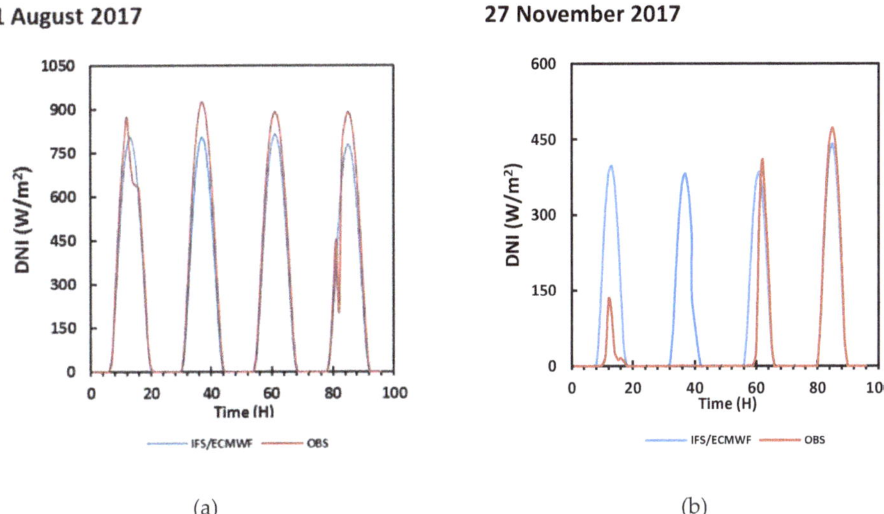

Figure 2. Example of four consecutive days of observed (red line) and forecasted (blue line) hourly mean DNI in Évora starting at (**a**) 1 August 2017 00:00 and (**b**) 27 November 2017 00:00.

The IFS/ECMWF forecasts have similar behavior to that of observational DNI for the two selected cases, with a fairly good agreement, especially in the case of August (Figure 2a). Figure 2b shows a partially cloudy day (day_0) and a cloudy day (day_1), making it evident that the model did not predict clouds correctly on November 28 since observational data clearly shows an overcast day. Another interesting feature in Figure 2 is that the IFS/ECMWF scheme slightly underestimated the DNI in the Summer case (Figure 2a) and overestimated it in the Autumn case (Figure 2b) in the case of a partly cloudy day. It is important to note that the example presented in Figure 2 is simply a selected example.

Figure 3 shows the comparison between ground-based measurements of hourly mean DNI and forecast data obtained from IFS/ECMWF for the entire period of study and the four forecasted days.

As expected, the errors associated with the hourly DNI forecast are quite significant with a strong scatter around y = x line (dashed line). The slope of the regression line indicates the quality of the forecasts and it is possible to conclude that the DNI IFS/ECMWF forecast is reasonable for the first three days ahead since the density of points is higher around the y = x line (dashed line in Figure 3). The worst forecast is for day_3. From Figure 3, it is also possible to verify that IFS overestimates DNI for lower values and underestimates DNI for higher irradiance values. This underestimation can be explained by the use of a constant monthly aerosol climatology in the IFS/ECMWF as argued by Lopes et al. [17], concluding that the model tends to underestimate DNI under very clear sky atmospheric conditions, when the actual aerosol concentrations are below mean values.

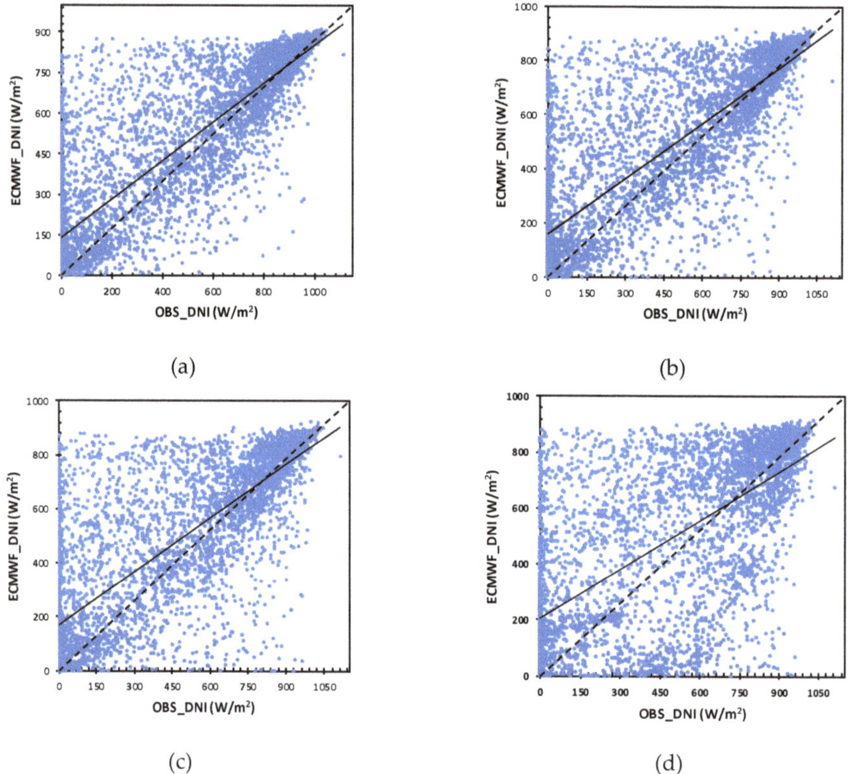

Figure 3. Scatter plots of predicted vs. measured hourly mean DNI for: (**a**) day_0, (**b**) day_1, (**c**) day_2, and (**d**) day_3, during the entire period considered. The dashed line represents the y = x line and the solid line is the least-squares regression fit.

The statistical errors, on an hourly basis, are presented in Table 1.

The assessment between datasets shows that errors increase from the first day of the forecast to the last day; day_0 exhibited the best performance with the lowest errors. MBE between calculated and measured DNI is smaller than 18W/m^2. Regarding the MAE and RMSE, their values increase from day_0 to day_3 (fourth day of the forecast) with a difference between them of ~61 W/m^2 (~45%) and ~72 W/m^2 (~37%), respectively. High correlation coefficients ($r \geq 0.70$) are obtained between the observations and forecasts for all forecast horizons (see Table 1).

The boxplots of Figure 4 show the MBE, MAE, RMSE, and correlation coefficient based on hourly values for each forecast day and the entire period of data (365 days).

MBE indicates a slight overestimation of hourly mean DNI for the majority of the forecast days (>50%) in the period. The length of the Interquartile Range (IQR) is a measure of the relative dispersion of a dataset and Figure 4a shows a similar length, in IQR, for the first two days of forecasts with values in (~−80; ~100 W/m^2). On the other hand, the difference between the IQR of the first forecast day and the last one (day_4) in the same plot is about 24%. Concerning MAE and RMSE, as expected, a similar pattern like MBE was found, with errors increasing as the lead time of the forecast increases, and a relative percentage error, relatively to the mean, between day_0 and day_3, for both parameters, of the order of 30%. It is worth noting that a significant number of outliers exist after the second day of forecasting. As for the correlation coefficients, these values indicate a good forecast performance with the best results obtained for day_0 with the highest median value of ~0.98 (Figure 4d). The correlation coefficient (r) presents a good performance for all forecast days in the analysis.

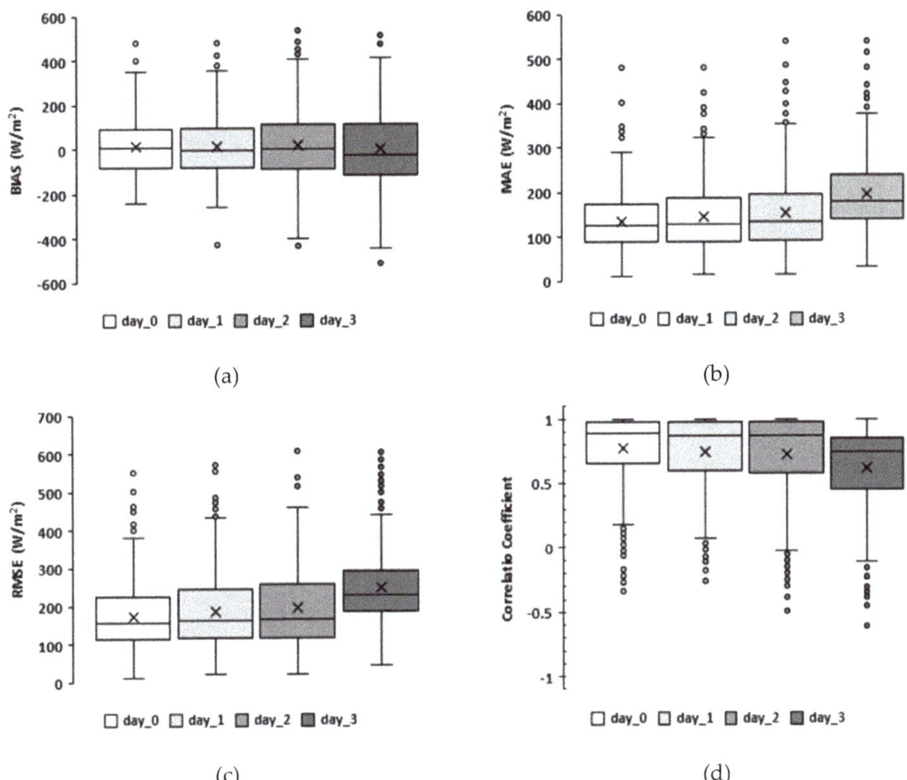

Figure 4. Boxplots of statistical indicators based on hourly values for (**a**) Mean Bias Error (MBE), (**b**) Mean Absolute Error (MAE), (**c**) Root Mean Square Error (RMSE), and (**d**) Correlation Coefficient (r), for different days ahead of forecast. The crosses represent the mean value of the sample, the horizontal solid line within the box represents the median, and the bottom and top of the boxes indicate the first and third quartiles, respectively. Boxes correspond to the Interquartile Range (IQR) where 50% of the data is located. The circles represent the outliers, and the lower and upper ends of the whiskers are the minimum and maximum values of the datasets, respectively.

Considering now the daily mean values, Figure 5 shows the comparison between measured and predicted DNI for the entire period.

As observed in the case of hourly values, the differences between observed and predicted DNI increases from the first day of the forecast to the last one. Another common feature observed is the DNI overestimation for lower values of direct normal irradiance as it can be seen through the trend lines. The statistical indicators obtained are comparable to the analysis made for the hourly values, showing a low forecast bias, with MBE values below 7 W/m^2 for all forecast days. Regarding RMSE, an increase of 35% percent (from ~61W/m^2 to 76 W/m^2) between the first and the last day of forecast. It is evident from the scatter plots of Figure 5 that between roughly 250 and 350 W/m^2 the distribution of data points is closer to the y = x line (ratio 1:1), which reveals a good agreement between observations and predictions. The overestimation occurs for observational DNI values below 200 W/m^2, with a larger dispersion, which may reflect inaccuracies in IFS cloud representation.

Figure 6 shows the monthly mean of daily values of simulated and measured DNI for the period between 1 August 2018 and 31 July 2019, thus allowing us to analyze the similarity between datasets throughout the year for the different forecast days.

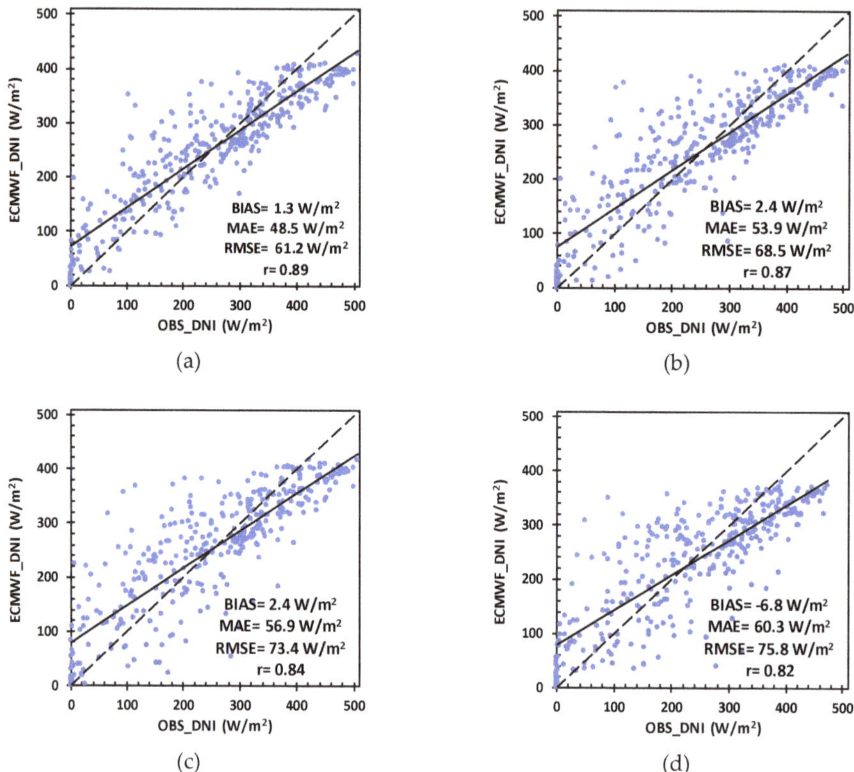

Figure 5. Comparison between predicted and measured daily mean DNI for the four prediction days: (**a**) day_0, (**b**) day_1; (**c**) day_2; (**d**) day_3. MBE, MAE, RMSE, and r are also presented in each plot. The solid lines are the linear fits and the dashed line represents the y = x line.

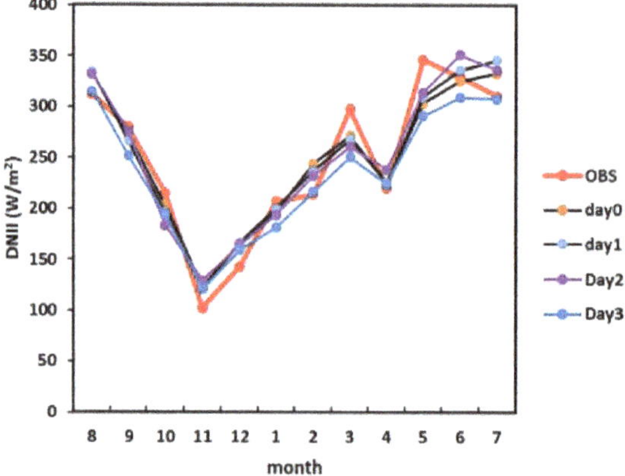

Figure 6. Monthly mean of predicted and observed daily mean DNI in Évora for the four different forecast days in the period from August 2018 to July 2019.

As shown in Figure 6, the IFS/ECMWF model overestimates the radiation in more ~50% of the days throughout the year, independently of forecast day, although with small differences datasets.

The variation of statistical indicators between datasets (daily) grouped by months is presented in Figure 7.

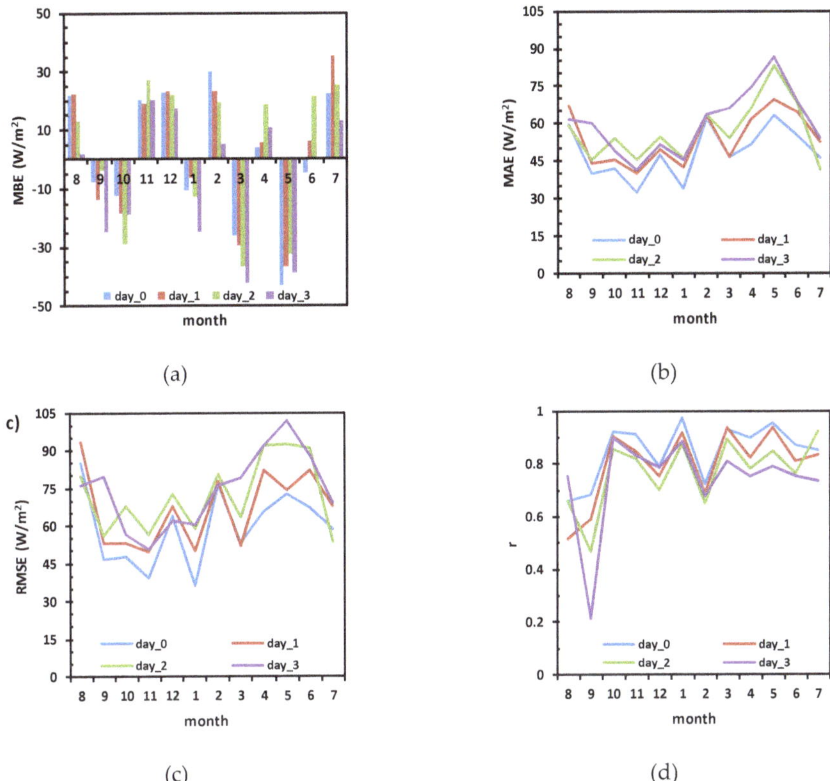

(a) (b) (c) (d)

Figure 7. Statistical indicators obtained from the comparison between measurements and predictions of daily mean DNI values (**a**) MBE; (**b**) MAE; (**c**) RMSE; (**d**) correlation coefficient.

Overall, MBE, MAE, and RMSE present better results for the first day of the forecast (day_0). The highest values of statistical errors correspond to the forecasts obtained for day_3.

From Figure 7a, MBE values range from about −42 W/m^2 to 35 W/m^2, and show ~60% of the months with positive MBE values (independently of forecast days). According to the same figure, the underestimation occurs in two-thirds of the months belonging to the MMA and SON seasons, probably as a consequence of a less accurate representation of the clouds (or aerosols) at short time scales in the radiative scheme of IFS/ECMWF.

According to Lopes et al. [17], the IFS global model from ECMWF tends to underestimate DNI in clear sky conditions due to the use of a monthly mean profile of aerosols. Perdigão et al. [6] also used the same argument in the assessment and characterization of the shortwave downward radiation incident at the Earth's surface over the Iberian Peninsula using the mesoscale Weather Research and Forecasting (WRF) model.

Figure 7b,c shows that MAE and RMSE present lower values, independently of forecast day, in JJA and SON seasons. MAE and RMSE present a similar variation that as found for MBE (Figure 7b,c) with values ranging from 33 W/m^2 to 87 W/m^2 and 36 W/m^2 to 102 W/m^2, respectively. As mentioned above, results show high correlation coefficients for all forecast days, although with the IFS/ECMWF model performing better on the first day of forecast.

The majority of statistical errors found in this work are in line with values obtained in the forecast of DNI by Nonnenmacher et al. [43] and Lara-Fanego et al. [18] using WRF model, and Gala et al. [44] using a clear sky model, among other studies. Table 2 of the article of Law et al. [13], show a summary of the state-of-the-art of DNI accuracy obtained from NWP and other methodologies.

3.2. Relation between the DNI Attenuation Index (DAI) and DNI Forecasts

Inaccurate representation of clouds in the radiative transfer scheme of global numerical weather prediction models is the primary cause of errors in the prediction of solar radiation. In this section, the DNI attenuation index (DAI) is proposed to assess and analyze the impact of cloud representation in the DNI forecast from the IFS/ECMWF model, as defined in Section 2.4 (Equation (6)).

Before analyzing the relationship between the DAI index, computed using data from the Évora radiometric station (Section 2.1) and the quality of the DNI forecast errors, the reliability of DAI is assessed, monthly, using the Total Cloud Area Fraction from CERES (Section 2.4) for the same local. According to Almorox et al. [45], global solar irradiation obtained from CERES, monthly, provides very good accuracy for solar radiation studies since their results show a good fit between CERES data and solar radiation data from different meteorological stations over Spain.

The linear regression between CERES cloud fraction and DAI shows a good agreement between these two indexes (Figure 8a), with a correlation coefficient of r ~0.92. When comparing the temporal evolution of DAI and cloud area fraction (CERES) monthly, both time series exhibits a similar pattern (Figure 8b), and shows, as expected, a decrease of cloud cover in the JJA season in contrast with an increase of the cloud cover in the DJF season.

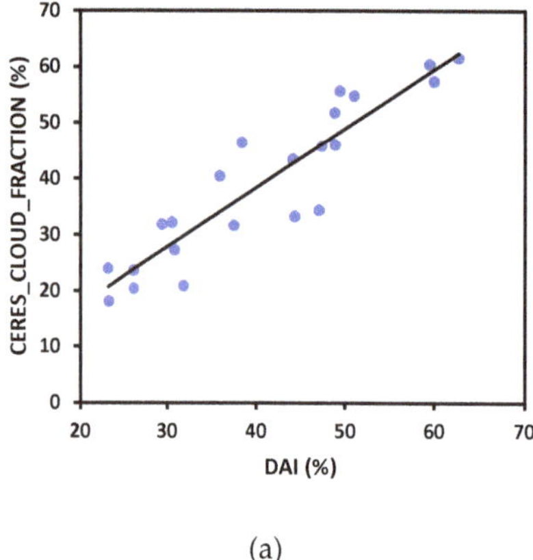

(a)

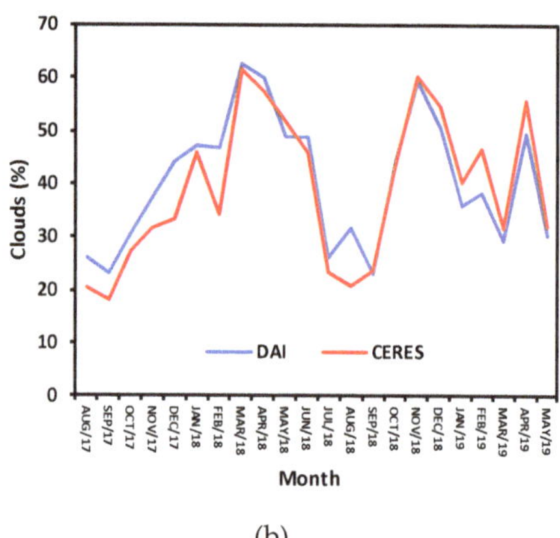

(b)

Figure 8. (**a**) Monthly mean cloud cover from Clouds and the Earth's Radiant Energy System (CERES) versus DNI Attenuation Index (DAI) in Evora, and (**b**) temporal evolution of CERES cloud fraction (red line) and DAI (blue line) in the period between August 2017 and July 2019 (twenty-two months). The black solid line represents the linear fit.

The major discrepancies between datasets occur in February 2018, February 2019, and in August 2018, corresponding to periods in which the region was affected by aerosol events (Saharan dust particles, in February, and forest fires in August).

The results suggest that DAI may be used as a proxy to cloud cover, particularly suitable to estimate the impact of clouds on the DNI forecast.

As for the observations, the DAI of model predictions was also calculated and hereafter is referred to as DAI (IFS). Figure 9 shows a boxplot comparison between DAI and DAI (IFS) index.

From Figure 9 it can be seen that in the majority of analyzed months, the DAI (IFS) is lower than DAI, meaning that the cloud scheme in IFS/ECMWF model underestimates the clouds and aerosols events when compared with the DAI index. DAIs are characterized by a lower variability in summer with more than 50% of the days with values lower than 31.25% (clear sky days) and a higher variability during the spring season (higher IQR values with more than 50% of days with values higher than 31.25%). These results are in line with the study by Roye et al. [46], in which low levels of cloudiness (clear skies days) over the Iberian Peninsula were found for the case of summer months, using satellite data from the MODIS and for the period 2001–2017, except for the Cantabrian coast. On the other hand, the variability of observed DAI is higher, mainly due to a higher variability on actual cloud cover and aerosol concentrations values. This variability also explains the relative high errors reported in the previous section. Perdigão et al. [6] also found high variability in downward shortwave radiation during March.

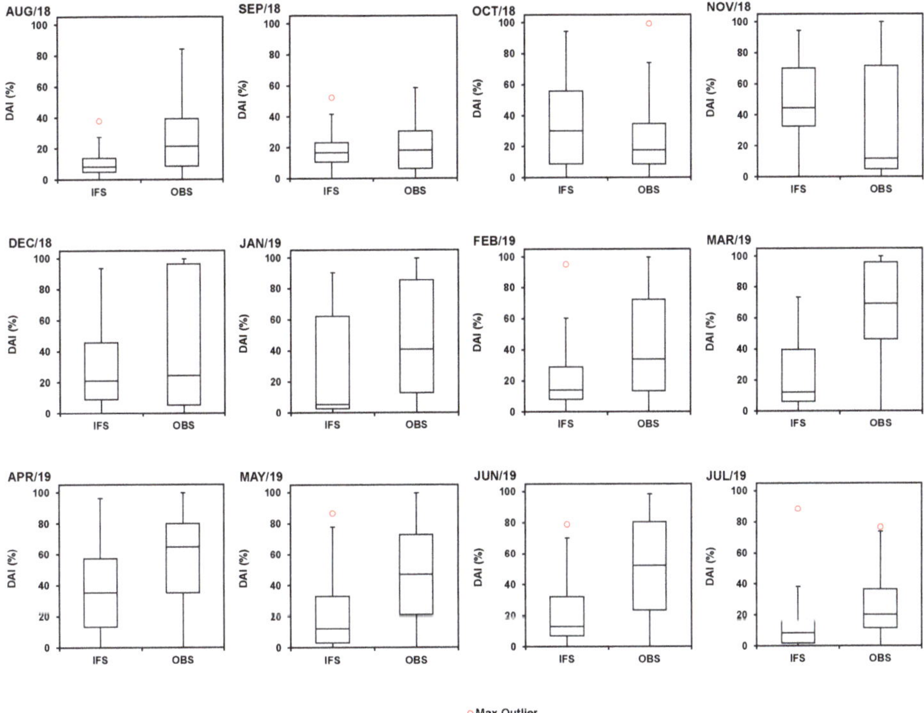

Figure 9. Monthly boxplots of daily mean values of DAI based on observations (OBS) and IFS/ECMWF forecasts (IFS), for Évora. The red circles represent outliers (maximum value).

Figure 10 shows the relation between the cloud class, grouped according to cloud coverage (in oktas), estimated based on the DAI, as indicated in Section 2.4, and the statistical indicator RMSE (in three ranges of values), obtained for each day, from hourly raw values. Day type class (I-clear, II-partially cloudy, and III-overcast) is obtained following the WMO guidelines [23].

From both plots of Figure 10, it becomes evident that RMSE strongly depends on the sky conditions, and it is possible to verify that:

(i) Approximately ~19% of the days present RMSE values lower than 100 W/m^2 (blue dots in Figure 10a). This percentage corresponds mostly to a cloud coverage lower than or equal to two oktas-clear skies days;

(ii) ~47% of the days present a cloud coverage of class II type, in the range (100–200 W/m^2);

(iii) RMSE values above 200 W/m^2 occur for ~34% of the days (red dots in Figure 10a). For this value, the majority of days are found in the cloud coverage type II category, suggesting that the model gives worst results in partially cloudy days, due to an inaccurate cloud representation or of their effects on the solar irradiance at the surface. For instance, Lopes [47] found that thin clouds (like cirrus) may cause a decrease in DNI of around 20%;

(iv) The errors found in summer months can be explained by the monthly constant aerosol climatology used in IFS/ECMWF as argued by Lopes et al. [17].

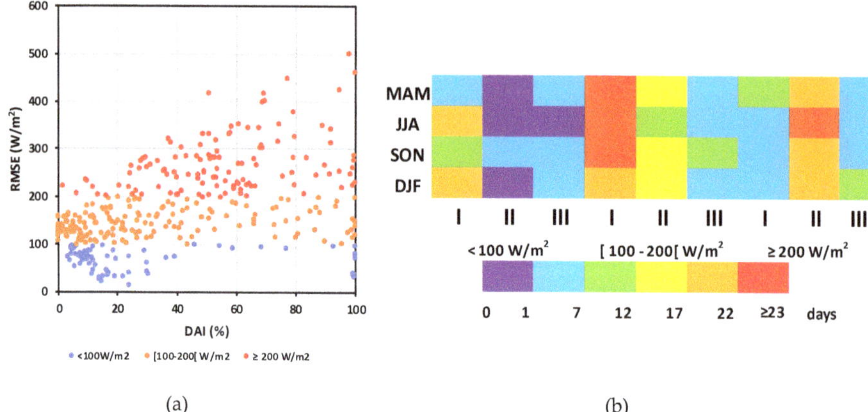

(a) (b)

Figure 10. (**a**) Scatter plot of DAI versus Root Mean Square Error (RMSE) for day_0 and (**b**) the number of days in a seasonal basis within different ranges of forecast errors grouped in classes, according to the cloud coverage–class I (0–2 oktas), class II (3–5 oktas), or class III (6–8 oktas) for RMSE.

The analysis of the climatology of cloud cover at Évora based on DAI for the period from 1 August/2018 to 31 July 2019 (Figure 10b), shows that about 45% of the days are in class I, considering only RMSE errors below 200 W/m^2, and these days are mainly in the MMA and JJA seasons, when more clear skies days occur over Évora city. These values are consistent with those found by Sanchez-Lorenzo et al. [48] and by Perdigão et al. [6] for the same sky conditions over the Iberian Peninsula.

Concerning the cloud coverage of class II and III, for the period in analysis and independently of the RMSE values, there are ~37% and ~13% of days, respectively. Cloud coverage of type III is mostly found in the DJF season.

It is important to mention that the quality and reliability of the forecast of solar radiation are directly related to the accuracy of cloud representation as well as aerosols. For instance, Lara-Fanego et al. [18] found that RMSE values ranged from 20% to 100% for clear and cloudy skies, respectively, using the WRF model over Andalusia.

Another feature can be seen in Figure 11 where it is possible to observe a relation between the RSR and the DAI for the first day of forecast (day_0).

As expected, from Figure 11, as the DAI increases the RSR increase. The forecast obtained from IFS/ECMWF presents a better performance for clear sky days. In general, the DNI forecast tends to be more accurate (RSR <0.5) for lower values of DAI, which represents ~50% of the days in Évora. This value of clear sky days is of the same order of the study of Freile-Aranda et al. [49], who found, for a climatic region which includes Évora, a minimum cloud cover (at an annual average) around 43%. The best performance (of IFS/ECMWF) are found for summer months (smaller values of RSR ≤0.5 and DAI ≤2 oktas). For instance, Kraas et al. [12], also found that during the summer season the forecasts are generally more reliable than in other seasons.

3.3. Statistical Bias Correction Analysis of Daily DNI Forecasts

The bias between solar radiation forecasts from Numerical Weather Prediction models and observations can be decreased by applying a bias post-processing correction. This Bias Correction (BC) methodology have been used in solar radiation studies by several authors, such as Ruiz-Arias et al. [7], Polo et al. [40,49], Perdigão et al. [6], Mejia et al. [41], among other authors.

The forecast values of DNI are corrected following the methodology described in Section 2.5. In Figure 12, the show the result of the application of the MOS method for the various forecast horizons.

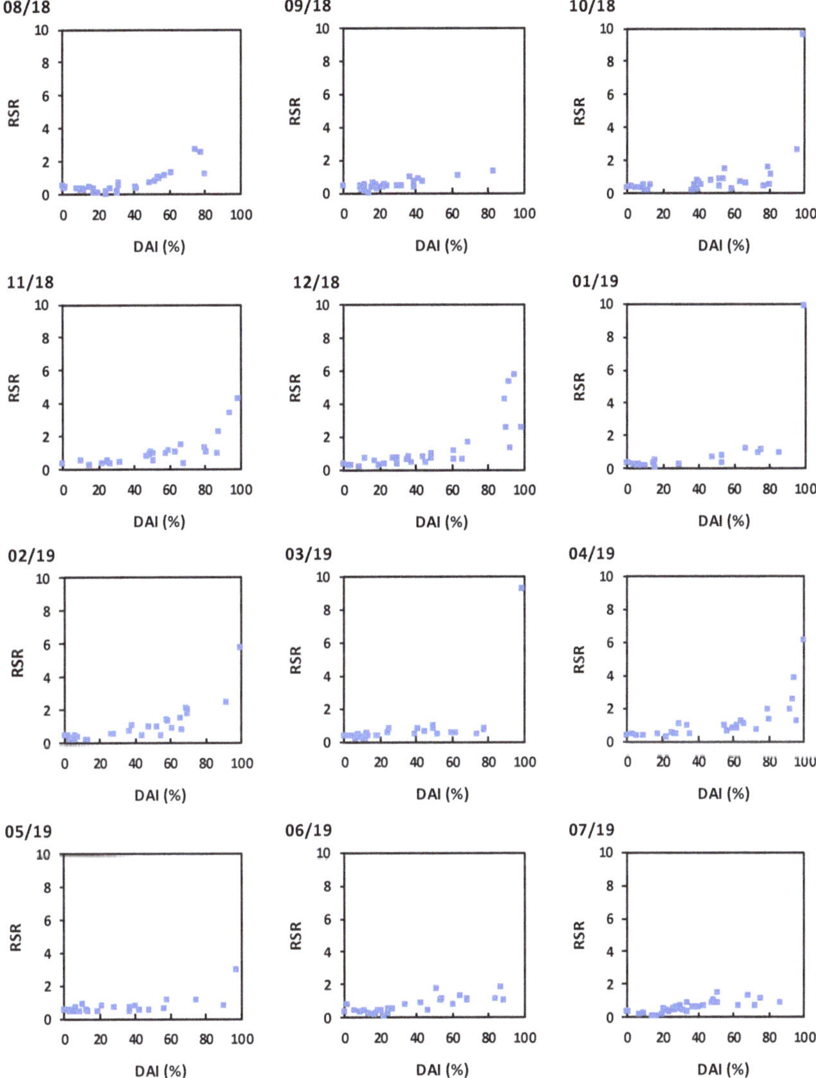

Figure 11. Relation between DAI and RMSE-observations standard deviation ratio (RSR) RSR based on hourly mean DNI forecasts and measurements for the first day of predictions between 1 August 2018 and 31 July 2019 (one year). RSR is dimensionless varying between zero and a large positive number.

Error metrics are also presented in graphs using the new corrected predictions.

Overall, MOS correction significantly improves the results. The corrected data exhibited (now) less dispersion around y = x line. Statistical errors decrease in the order of about 30% when compared with initial IFS predictions and independently of the forecast day. For example, MAE values decrease from the interval (49–60 W/m^2) to (32–41 W/m^2), while RMSE values decrease from (61–76 W/m^2) to (43–57 W/m^2). The correlation coefficient is in line with previous values, i.e., all *r* values were improved with values ≥0.89.

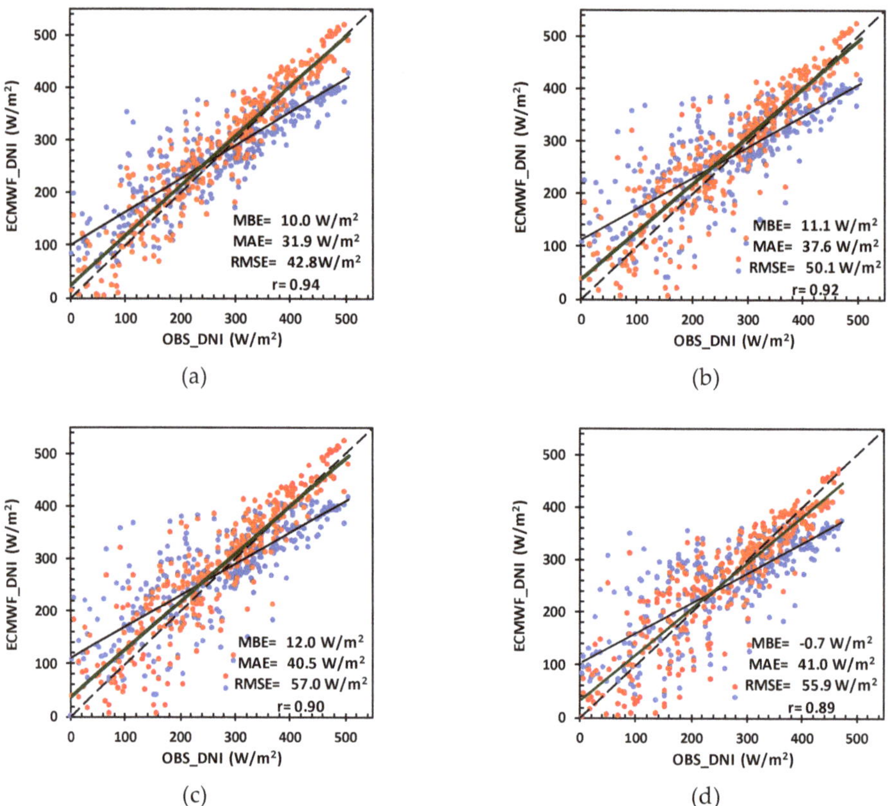

Figure 12. Comparison between predicted and measured daily mean DNI for the four prediction days: (**a**) day_0, (**b**) day_1; (**c**) day_2; (**d**) day_3 before (blue dots) and after Bias correction (red dots). MBE, MAE, RMSE, and r, after BC, are also presented in each plot. The solid lines are the linear fits-green for BC procedure and black for IFS/ECMWF raw data-and the dashed line represents the y = x line.

To a better comparison, Figure 13 shows the cumulative distribution functions (CDF) of daily DNI, for each day of forecast, before and after the correction procedure.

Results in Figure 13 show the similarity of the CDF between observational and IFS/ECMWF outputs before and after the bias correction. In general, and independently of the forecast day, as said before, the linear regression method successfully improved the DNI outputs with the new corrected cumulative distribution function plots closer to the observed DNI.

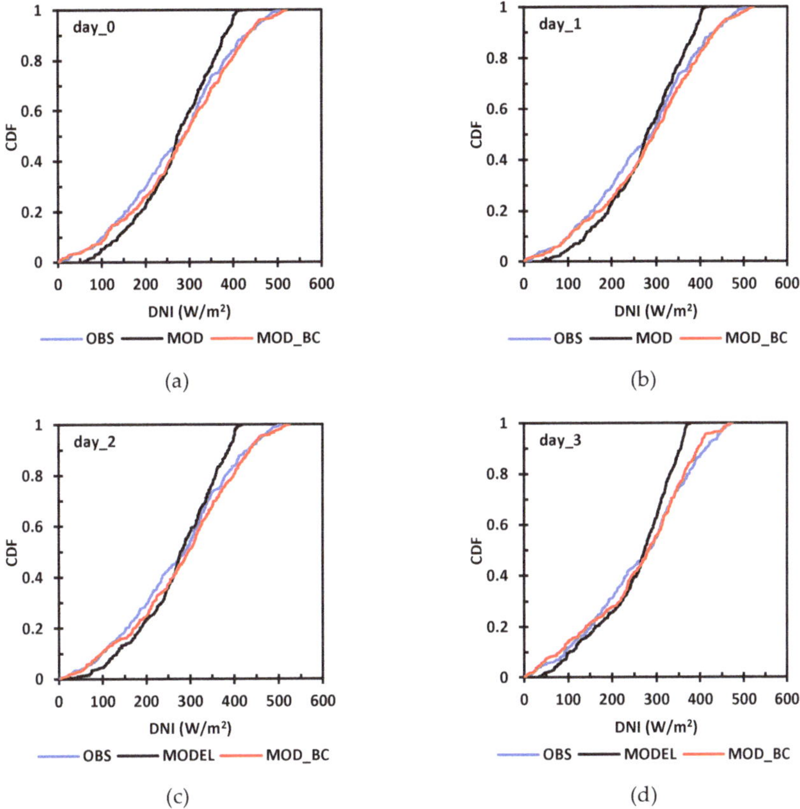

Figure 13. Cumulative distribution functions (CDF) of daily mean DNI grouped by day of forecast, (**a**) Day_0; (**b**) day_1; (**c**) day_2; (**d**) day_3, from 1 August 2018 to 31 July 2019, original forecasts (black line), forecasts after Bias Correction (red line) and observations (blue line).

4. Conclusions

Given the importance that alternative energies have in a sustainable economic, social, and environmental perspective, it is important to know in advance how solar, wind, or other renewable energy resources change on an hourly, daily, or monthly basis. In this work, DNI from Integrated Forecasting System of European Centre for Medium-Range Weather Forecasts (IFS/ECMWF) dataset was evaluated over one year (1 August 2018 to 31 July 2019) against observed DNI, at hourly time scales, for one station located at Evora (south of Portugal) for different days ahead of forecast (until three days ahead).

Statistical BIAS, MAE, RMSE, RSR, and correlation coefficient were used to assess the relation between IFS/ECMWF and observational DNI. Additionally, this paper, also describes a new methodology based on DNI observations (the called of DNI attenuation index–DAI) to estimate the transparency of the atmosphere in a particular region. The DAI was evaluated with total cloud cover parameter obtained from CERES product data and results showed high correlation coefficients between datasets, suggesting that DAI can be used as proxy to classify the cloud coverage in direct solar radiation studies. This index was used to analyze the relation between the cloud coverage and the predicted DNI, as well as the respective associated error.

The IFS/ECMWF DNI forecasts present similar magnitudes and pattern relatively to observational data, but the errors increase with the forecast lead time (from 1 to 3 days ahead). Discrepancies between modelled and measured radiation are relatively small mainly for the first three days of forecasts. The analysis of hourly data showed that the DNI is overestimated by IFS/ECMWF. Regarding the correlation coefficient, values are found above ≥0.7, independently of the forecast day. This work also shows that the first day ahead forecast (day_1) has similar error magnitudes in relation to the first 24 h forecast (day_0).

Hourly analysis also shows values of MBE, lower than 20 W/m^2. Regarding the MAE and RMSE values, an increase from day_0 to day_3 was observed, with a difference between the first and the third day of forecast ~45% and ~37%, respectively. High correlation coefficients (r ≥0.7) are found for all forecast days.

Daily analysis shows better results, with MBE values lower than 7 W/m^2 for all forecast days. In the case of RMSE, values increase about 35% percent from the first day of forecast to the last one. The correlation coefficients of daily data are higher than in the case of hourly data, ranging between 0.82 (day_3) and 0.89 (day_0).

The mean-monthly cloud coverage is well captured by DAI along the year. As expected, the observed DNI is higher in spring and summer months with the lowest values in DAI for the same seasons. The underestimation of cloud cover by the IFS/ECMWF seems to be evident since comparison between observed and predicted DAI reveals that model tends to underestimate the effects of clouds on DNI. This relation was also found for Andalusia (located in Iberian Peninsula) using WRF model by Lara-Fanego et al. [18] in the case of three days ahead DNI forecasts.

The accuracy of IFS/ECMWF to forecast DNI is higher for clear or partially cloudy sky days. DAI index confirms that the performance of the IFS/model decrease with an increasing of clouds/aerosols effects.

A bias correction post-processing through a linear regression was used to correct the IFS/ECMWF predictions, which has shown to significantly improve the forecast for Évora with a decrease in the order of 30% for all statistical error metrics, except for the correlation coefficient, independently of the days ahead in consideration.

The results obtained in this work are consistent with those obtained by Lopes et al. [17] for the same location, and by Nonnenmacher et al. [14], Troccoli, and Morcrette [19], among others, where it was found that errors increase with the lead time forecast. Overall, ECMWF DNI forecasts provide valuable information for the management and operation of CSP plants, especially after the usage of the post-processing bias correction.

Author Contributions: Conceptualization, J.P., P.C., R.S. and M.J.C.; formal analysis, J.P.; methodology, J.P., P.C., R.S. and M.J.C.; resources, P.C., R.S. and M.J.C.; supervision, P.C., R.S. and M.J.C.; writing—original draft, J.P.; writing—review and editing, J.P., P.C., R.S. and M.J.C. All authors have read and agreed to the published version of the manuscript.

Funding: The work is co-funded by the European Union through the European Regional Development Fund, included in the COMPETE 2020 (Operational Program Competitiveness and Internationalization) through the ICT project (UIDB/04683/2020) with the reference POCI-01-0145-FEDER-007690 and also through the DNI-A (ALT20-03-0145-FEDER-000011) and LEADING (PTDC/CTA-MET/28914/2017) projects.

Acknowledgments: The authors acknowledge the ECMWF for providing DNI forecasts through the website and at NCEP Reanalysis data provided by the NOAA/OAR/ESRL PSD, Boulder, Colorado, USA, from their Web site at https://www.esrl.noaa.gov/psd/. The authors also thank the CERES Science Team for the data provided.

Conflicts of Interest: The authors declare there is no conflict of interest with regard to this manuscript.

References

1. Gaetani, M.; Huld, T.; Vignati, E.; Monforti-Ferrario, F.; Dosio, A.; Raes, F. The near future availability of photovoltaic energy in Europe and Africa in climate-aerosol modeling experiments. *Renew. Sustain. Energy Rev.* **2014**, *38*, 706–716. [CrossRef]
2. Escobar, R.A.; Cortés, C.; Pino, A.; Salgado, M.; Pereira, E.B.; Martins, F.R.; Boland, J.; Cardemil, J.M. Estimating the potential for solar energy utilization in Chile by satellite-derived data and ground station measurements. *Sol. Energy* **2015**, *121*, 139–151. [CrossRef]

3. Santos, J.A.; Rochinha, C.; Liberato, M.L.R.; Reyers, M.; Pinto, J.G. Projected changes in wind energy potentials over Iberia. *Renew. Energy* **2015**, *75*, 68–80. [CrossRef]
4. Du, H.; Jones, P.J.; Lannon, S.C. Creating localised near future weather data for predicting the performance of buildings in the UK. In Proceedings of the 12th REHVA World Congress CLIMA 2016, Aalborg, Denmark, 22–25 May 2016; Heiselberg, P.K., Ed.; Department of Civil Engineering, Aalborg University: Aalborg, Denmark, 2016; p. 537.
5. Ruiz-Arias, J.A.; Gueymard, C.A.; Dudhia, J.; Pozo-Vazquez, D. Improvement of the weather research and forecasting (WRF) model for solar resource assessments and forecasts under clear skies. In Proceedings of the World Renewable Energy Forum, Denver, CO, USA, 13–17 May 2012.
6. Perdigão, J.; Salgado, R.; Magarreiro, C.; Soares, P.M.M.; Costa, M.J.; Dasari, H. An Iberian climatology of solar radiation obtained from WRF regional climate simulations for 1950–2010 period. *Atmos. Res.* **2017**. [CrossRef]
7. Ruiz-Arias, J.A.; Quesada-Ruiz, S.; Fernández, E.F.; Gueymard, C.A. Optimal combination of gridded and ground-observed solar radiation data for regional solar resource assessment. *Sol. Energy* **2015**, *112*, 411–424. [CrossRef]
8. Šúri, M.; Huld, T.A.; Dunlop, E.D.; Ossenbrink, H.A. Potential of solar electricity generation in the European Union member states and candidate countries. *Sol. Energy* **2007**, *81*, 1295–1305. [CrossRef]
9. Pereira, S.; Canhoto, P.; Salgado, R.; Costa, M.J. Development of an ANN based corrective algorithm of the operational ECMWF global horizontal irradiation forecasts. *Sol. Energy* **2019**, *185*, 387–405. [CrossRef]
10. Martín, L.; Zarzalejo, L.F.; Polo, J.; Navarro, A.; Marchante, R.; Cony, M. Prediction of global solar irradiance based on time series analysis: Application to solar thermal power plants energy production planning. *Sol. Energy* **2010**, *84*, 1772–1781. [CrossRef]
11. Alsamamra, H.; Ruiz-Arias, J.A.; Pozo-Vázquez, D.; Tovar-Pescador, J. A comparative study of ordinary and residual kriging techniques for mapping global solar radiation over southern Spain. *Agric. For. Meteorol.* **2009**, *149*, 1343–1357. [CrossRef]
12. Kraas, B.; Schroedter-Homscheidt, M.; Madlener, R. Economic Merits of a State-of-the-Art Concentrating Solar Power Forecasting System for Participation in the Spanish Electricity Market. *Sol. Energy* **2013**, *93*, 244–255. [CrossRef]
13. Law, E.W.; Prasad, A.A.; Kay, M.; Taylor, R.A. Direct normal irradiance forecasting and its application to concentrated solar thermal output forecasting—A review. *Sol. Energy* **2014**, *108*, 287–307. [CrossRef]
14. Nonnenmacher, L.; Kaur, A.; Coimbra Coimbra, C. Day-ahead resource forecasting for concentrated solar power integration. *Renew. Energy* **2016**, *86*, 866–876. Available online: http://www.sciencedirect.com/science/article/pii/S0960148115302688 (accessed on 3 December 2019). [CrossRef]
15. Gomez-Gil, F.J.; Wang, X.; Barnett, A. Analysis and prediction of energy production in concentrating photovoltaic (CPV) installations. *Energies* **2012**, *5*, 770–789. [CrossRef]
16. Casado-Rubio, J.; Revuelta, M.; Postigo, M.; Martínez-Marco, I.; Yagüe, C.A. Postprocessing Methodology for Direct Normal Irradiance Forecasting Using Cloud Information and Aerosol Load Forecasts. *J. Appl. Meteorol. Climatol.* **2017**, *56*, 1595–1608. [CrossRef]
17. Lopes, F.; Silva, H.; Salgado, R.; Cavaco, A.; Canhoto, P.; Colarres-Pereira, M. Short-term forecasts of GHI and DNI for solar energy systems operation: Assessment of the ECMWF integrated forecasting system in southern Portugal. *Sol. Energy* **2018**, *170*, 14–30. [CrossRef]
18. Lara-Fanego, V.; Ruiz-Arias, J.A.; Pozo-Vázquez, A.D.; Gueymard, C.A.; Tovar-Pescador, J. Evaluation of DNI forecast based on the WRF mesoscale atmospheric model for CPV applications. *AIP Conf. Proc.-Am. Inst. Phys.* **2012**, *1477*, 317. [CrossRef]
19. Troccoli, A.; Morcrette, J.J. Skill of direct solar radiation predicted by the ECMWF global atmospheric model over Australia. *J. Appl. Meteorol. Climatol.* **2014**, *53*, 2571–2588. [CrossRef]
20. Vick, B.D.; Myers, D.R.; Boyson, W.E. Using direct normal irradiance models and utility electrical loading to assess benefit of a concentrating solar power plant. *Sol. Energy* **2012**, *86*, 3519–3530. [CrossRef]
21. Haiden, T.; Rodwell, M.J.; Richardson, D.S.; Okagaki, A.; Robinson, T.; Hewson, T. Intercomparison of global model precipitation forecast skill in 2010/11 using the SEEPS score. *Mon. Weather Rev.* **2012**, *140*, 2720–2733. [CrossRef]
22. Kipp & Zonen. Available online: http://www.kippzonen.com/ (accessed on 15 September 2019).

23. World Meteorological Organization (WMO). *Guide to Meteorological Instruments and Methods of Observation, (WMO-No. 8)*, 7th ed.; World Meteorological Organization (WMO): Geneva, Switzerland, 2008; ISBN 978-92-63-100085.
24. ISO. *ISO 9060:1990: Specification and Classification of Instruments for Measuring Hemispherical Solar and Direct Solar Radiation*; ISO: Geneva, Switzerland, 1990.
25. Long, C.N.; Dutton, E.G. BSRN Global Network recommended QC tests, V2. *J. Clim.* **2010**, *25*, 8542–8567. [CrossRef]
26. Hogan, R.J.; Bozzo, A. ECRAD: A new radiation scheme for the IFS. *ECMWF Tech. Memo.* **2016**, *787*, 33p.
27. Morcrette, J.-J.; Barker, H.; Cole, J.; Iacono, M.; Pincus, R. Impact of a new radiation package, mcrad, in the ecmwf integrated forecasting system. *Mon. Weather Rev* **2008**, *136*, 4773–4798. [CrossRef]
28. Tegen, I.; Hoorig, P.; Chin, M.; Fung, I.; Jacob, D.; Penner, J. Contribution of different aerosol species to the global aerosol extinction optical thickness: Estimates from model results. *J. Geophys. Res.* **1997**, *102*, 23895–23915. [CrossRef]
29. Flemming, J.; Benedetti, A.; Inness, A.; Engelen, R.J.; Jones, L.; Huijnen, V.; Rémy, S.; Parrington, M.; Suttie, M.; Bozzo, A.; et al. The CAMS interim reanalysis of carbon monoxide, ozone and aerosol for 2003–2015. *Atmos. Chem. Phys.* **2017**, *17*, 1945–1983. [CrossRef]
30. Fu, Q. An accurate parameterization of the solar radiative properties of cirrus clouds. *J. Clim.* **1996**, *9*, 2058–2082. [CrossRef]
31. European Center for Medium Time Weather Forecasting (ECMWF). Available online: www.ecmwf.int (accessed on 25 August 2019).
32. Wild, M.; Schmucki, E. Assessment of global dimming and brightening in IPCC-AR4/CMIP3 models and ERA-40. *Clim. Dyn.* **2011**, *37*, 1671–1688. [CrossRef]
33. Perez, R.; Lorenz, E.; Pelland, S.; Beauharnois, M.; Van Knowe, G.; Hemker, K.; Heinemann, D.; Remund, J.; Müller, S.C.; Traunmüller, W.; et al. Comparison of numerical weather prediction solar irradiance forecasts in the US, Canada and Europe. *Sol. Energy* **2013**, *94*, 305–326, ISBN 0038-092X. [CrossRef]
34. Chai, T.; Draxler, R.R. Root mean square error (RMSE) or mean absolute error (MAE)?—Arguments against avoiding RMSE in the literature. *Geosci. Model Dev.* **2014**, *7*, 1247–1250. [CrossRef]
35. Landelius, T.; Lindskog, M.; Körnich, H.; Andersson, S. Short-range solar radiation forecasts over Sweden. *Adv. Sci. Res.* **2018**, *15*, 39–44. Available online: https://www.adv-sci-res.net/15/39/2018/asr-15-39-2018.pdf (accessed on 5 December 2019). [CrossRef]
36. Moriasi, D.N.; Arnold, J.G.; Van Liew, M.W.; Binger, R.L.; Harmel, R.D.; Veith, T. Model evaluation guidelines for systematic quantification of accuracy in watershed simulations. *Trans. ASABE* **2007**, *50*, 885–900. [CrossRef]
37. Iqbal, M. *An Introduction to Solar Radiation*; Academic Press: New York, NY, USA, 1975.
38. Jafariserajehlou, S.; Mei, L.; Vountas, M.; Rozanov, V.; Burrows, J.P.; Hollmann, R. A cloud identification algorithm over the Arctic for use with AATSR–SLSTR measurements. *Atmos. Meas. Tech.* **2019**, *12*, 1059–1076. [CrossRef]
39. Wielicki, B.A.; Barkstrom, B.R.; Harrison, E.F.; Lee, R.B., III; Louis Smith, G.; Cooper, J.E. Clouds and the Earth's Radiant Energy System (CERES): An earth observing system experiment. *Bull. Am. Meteorol. Soc.* **1996**, *77*, 853–868. [CrossRef]
40. Polo, J.; Martin, L.; Vindel, J.M. Correcting satellite derived DNI with systematic and seasonal deviations: Application to India. *Renew. Energy* **2015**, *80*, 238–243. [CrossRef]
41. Mejia, J.F.; Giordano, M.; Wilcox, E. Conditional summertime day-ahead solar irradiance forecast. *Sol. Energy* **2018**, *163*, 610–622. [CrossRef]
42. Polo, J.; Wilbert, S.; Ruiz-Arias, J.A.; Meyer, R.; Gueymard, C.; Súri, M.; Martín, L.; Mieslinger, T.; Blanc, P.; Grant, I.; et al. Preliminary survey on site-adaptation techniques for satellite-derived and reanalysis solar radiation datasets. *Sol. Energy* **2016**, *132*, 25–37. [CrossRef]
43. Nonnenmacher, L.; Kaur, A.; Coimbra, C.F.M. Verification of the SUNY direct normal irradiance model with ground measurements. *Sol. Energy* **2014**, *99*, 246–258. [CrossRef]
44. Gala, Y.; Fernandez, A.; Dıaz, J.; Dorronsoro, J. Support vector forecasting of solar radiation values. In *Hybrid Artificial Intelligent Systems*; Pan, J.-S., Polycarpou, M., Wozniak, M., Carvalho, A.P.L.F., Quintian, H., Corchado, E., Eds.; Springer: Berlin/Heidelberg, Germany, 2013; pp. 51–60.

45. Almorox, J.; Ovando, G.; Sayago, S.; Bocco, M. Assessment of surface solar irradiance retrieved by CERES. *Int. J. Remote Sens.* **2017**, *38*, 3669–3683. [CrossRef]
46. Royé, D.; Lorenzo, N.; Rasilla, D.; Martí, A. Spatio-temporal variations of cloud fraction based on circulation types in the Iberian Peninsula. *Int. J. Climatol.* **2019**, *39*, 1716–1732. [CrossRef]
47. Lopes, M. Desenvolvimento de um Sistema de Baixo custo para a Previsão da Irradiância Solar a Curto Prazo. Master's Thesis, Instituto Superior Técnico, Lisbon, Portugal, 2015. Available online: https://fenix.tecnico.ulisboa.pt/downloadFile/281870113702447/Dissertacao_67932 (accessed on 3 December 2019).
48. Sanchez-Lorenzo, A.; Calbó, J.; Brunetti, M.; Deser, C. Dimming/brightening over the Iberian Peninsula: Trends in sunshine duration and cloud cover and their relations with atmospheric circulation. *J. Geophys. Res.* **2009**, *114*, D00D09. [CrossRef]
49. Freile-Aranda, M.D.; Gomez-Amo, J.L.; Utrillas, M.P.; Pedros, R.; Martínez-Lozano, J.A. Seasonal analysis of cloud characteristics and radiative effect over the Iberian Peninsula using MODIS-CERES observations. *Tethys* **2017**, *14*, 3–9. [CrossRef]

Article

Performance Comparison between Deep Learning and Optical Flow-Based Techniques for Nowcast Precipitation from Radar Images

Marino Marrocu * and Luca Massidda

CRS4, Center for Advanced Studies, Research and Development in Sardinia, loc. Piscina Manna ed. 1, 09050 Pula, Italy; luca.massidda@crs4.it

* Correspondence: marino.marrocu@crs4.it

Received: 26 May 2020; Accepted: 22 June 2020; Published: 24 June 2020

Abstract: In this article, a nowcasting technique for meteorological radar images based on a generative neural network is presented. This technique's performance is compared with state-of-the-art optical flow procedures. Both methods have been validated using a public domain data set of radar images, covering an area of about 10^4 km^2 over Japan, and a period of five years with a sampling frequency of five minutes. The performance of the neural network, trained with three of the five years of data, forecasts with a time horizon of up to one hour, evaluated over one year of the data, proved to be significantly better than those obtained with the techniques currently in use.

Keywords: nowcast; meteorological radar data; optical flow; deep learning

1. Introduction

Forecasting precipitation at very high resolutions in space and time, particularly with the aim of providing the boundary conditions for hydrological models and estimating the associated risk, is one of the most difficult challenges in meteorology [1]. In short, the problem is to predict the field of precipitation on grids of 1 km or less and for horizons of less than a few hours (nowcast) [2]. Due to the high spatial resolutions required, methods based on meteorological models are not effective, because they are too onerous computationally and the time to perform a simulation is usually excessively too long for operational purposes. They also depend, in a critical way, on the initial conditions as the level of uncertainty is too high and it is difficult to assimilate locally recorded data [3,4]. Even if these difficulties were overcome, a fundamental difficulty would remain, the gap between the spatial scales provided by the meteorological models and those useful to hydrologists for operational purposes. To overcome this difficulty and quantify the uncertainty involved, probabilistic approaches were proposed [5].

In contrast, simplified approaches based on image processing and related data extrapolation, based on radar reflectivity or remote sensing data, have proven to be more effective. For example in [6], an application in Mediterranean basins of a semi-distributed hydrological model using rainfall estimated by radar was discussed, and in a more recent paper an application with a coupled soil and groundwater model was discussed [7]. A study discussing the propagation of the uncertainty in weather radar estimates of rainfall through hydrological models can be found in [8]. This approach showed to be even more promising by using new methodologies based on recent advances in deep learning techniques [9].

The simplest techniques for nowcasting on meteorological radar data consist of two phases, tracking and extrapolation, i.e., first the velocity field is estimated, typically with image comparison techniques called Optical Flow (OF), this field is then used to extrapolate a prediction of radar images through semi-Lagrangian schemes, or interpolation procedures [10]. This procedure estimates the precipitation as seen by the radar, assuming that the rate of advection evaluated by comparing a sequence of two or more consecutive images stays constant with no terms of sink or source altering the intensity of the field in the chosen forecast range. Unfortunately these two assumptions, given the high space-temporal variability of the precipitation, most never occur, and this is all the more true the longer the forecast time horizon is. For this reason, a forecast procedure based on optical flow has a limited validity with lead times of the order of the hour. Despite this, there are several nowcasting procedures for radar images and related by-products based on OF, operational in meteorological centers, that are able to provide predictions with appreciable skills with a time horizon of up to three hours [11,12], when used with additional information from numerical model's data, and real-time precipitation measurements. For this reason, the topic, although widely investigated, is still of interest for research [13]. This work will focus on the analysis of nowcast methods based only on radar data. The intent is to assess the degree of the predictive potential of the data itself and with the idea that this can add value even when operating in conjunction to additional information.

The first nowcasting methods based on OF [14] foresaw the evolution of the image from one instant to the next by means of a rigid translation, with an advection field determined by the criteria of maximum correlation between two or more images, immediately preceding the time of emission of the forecast. In the TREC method, the advection field is not constant but is evaluated by comparing boxes of pixels of equal size between two successive radar images, and identifying similar boxes through cross correlation. An advection vector connects the centers of two similar boxes. Repeating the process on boxes of optimal size distributed on the image, it is possible to determine a non constant field of advection that is then used for the extrapolation of the precipitation field [15]. Over the years these methods have been refined, providing the possibility to advect separately the features defined as "interesting" within the images, such as the position of the precipitating centers, or the maximum gradient points, as in the method described in Shi-Tomasi [16]. Then the possibility of deformation of the advected pattern was added, applying for example an affine transformation [17]. All these methods and several other variants have the conceptual limitation of being deterministic, not taking into account the high uncertainty inherent in this overly simplistic approach. The so-called SPROG [18] approach takes into account the diverse predictability of the precipitation at different scale. It decomposes the logarithmic transformation of the initial field into a multiplicative cascade of decreasing spatial scale fields, and advects each of them separately, keeping the coherence with the pre-forecast period. Finally, the field at each forecast deadline is calculated as the antilogarithm of the product of the fields at the individual scales. More details of this and other techniques used as benchmarks will be provided in the following section.

More recently, the potential offered by the highly non-linear approach of neural networks has been investigated. These techniques use a large amount of past data to "learn" and devise a predictive model. The first studies that have started this line of research for forecasting precipitation from radar images have been introduced by the Hong-Kong Meteorological Observatory Group [19,20]. In a work published in 2015 [19], the authors introduced the ConvLSTM model and demonstrated that it outperforms the OF-based operating model, called SWIRLS (Short term forecast of Intense Rain Localized Systems), for forecasting horizons of up to two hours. In a follow up [20], the same authors have introduced the TrajGRU model, which uses the same convolutional and recursive deep networks as the previous one, but where a location feature for radar image processing is introduced, improving the performance of both SWIRLS and convLSTM models. Recently a variant of PredNet architecture has been proposed in [21] while in [22] a completely convolutional deep neural network is used, investigating in particular the importance of data pre-processing, network structure and loss function.

With the aim of investigating and comparing the predictive potential of different approaches and, in order to find methods that can offer room for improvement, this work proposes and evaluates the performance of a PredNet deep neural network [23]. The architecture is borrowed from computer vision for short-term video prediction, where it represents one of the state of the art approaches.

Compared to [21], where a similar network was used, this version has a modified loss function for training phase that tries to maximize the predictive performance of the architecture. A quantitative and qualitative comparison of the results obtained with the PredNet versus the reference OF methods is presented here that will highlight not only the different performance but also the distinct characteristics of the forecast obtained, and their limits.

The rest of the paper is organized as follows: Section 2 illustrates the methodologies used as benchmarks, Section 3 presents the modified neural network used in this analysis, Section 4 describes the data used for the numerical comparison while results are discussed in Section 5. Section 6 hosts the conclusions and a discussion of potential future developments.

2. Benchmark Nowcast Techniques

The methodologies used as benchmarks to assess the performance of the approach proposed in this paper are those implemented in two public domain libraries called *rainymotion* [10] and *pysteps* [24].

The *rainymotion* library, built on *opencv* [25], implements four different nowcasting methods called Dense, Dense Rotation, SparseSD and Sparse. The Dense and Dense Rotation methods are based on standard Optical Flow estimation techniques. The Sparse and SparseSD methods, instead of tracking the displacement of all pixels in the image, first detect a chosen number of "good features to track" [16], typically the boundaries of moving objects with respect to the constant background, and trace their most likely displacement with optical flow techniques [14]. Next, they calculate, and apply, the affine transformation that maps the "interesting" features from one frame to the following [17] and eventually fill the final image by interpolation [26]. The Sparse method uses a sequence of images of variable length (in our case 12) preceding the instant of the nowcast, in order to trace the "interesting" characteristics of the field, while SparseSD uses only the last two images. Since, in all the tests we carried out, the Sparse (SP henceforth) and Dense Rotation (DR henceforth) methods have proven to perform better than the corresponding SparseSD and Dense methods, the results presented below will be only related to SP and DR.

Pysteps [24] is also a public domain library, managed and developed by a large community, which besides being an operational tool for nowcasting, is used as a development and study platform for the implementation of new techniques. For the estimation of the field of advection, pysteps implements three methodologies: a local type based on the classic Lucas-Kanade (LK) technique [14,27], a global variational type (VET) [28,29] and a spectral type (DARTS) [30]. Extrapolation by advection is done with a semi-lagrangian scheme backward in time [29]. The most sophisticated technique of the pysteps library, used here as a benchmark, is called STEPS, takes into account the different predictability of the precipitation phenomena at different scales. In practice, using an approach called SPROG [18], after a logarithmic transformation, a decomposition of the initial field into a multiplicative cascade of fields with decreasing spatial scales is performed and each of these spatial fields is advanced with an advection obtained consistently during the period before the nowcast. The decomposition of the field at different scales can be achieved using techniques of varying degrees of complexity, such as Discrete Cosine Transform or wavelet. In this case an FFT was used in which the different portions of the spectrum are weighed by Gaussian functions [31] and then, by inverse FFT, brought back into the spatial domain. The additional uncertainty due to the fact that, during the period of the forecast validity there may be growth, decay, creation and/or definitive disappearance of the precipitating cells, alongside the uncertainties linked to different spatial scales, were taken into account trough a stochastic noise built on the space and time variability, observed in the period preceding the emission, added to each partial field. For the implementation details please refer to [24] and the references therein. This approach makes it possible to obtain not a single deterministic forecast, but a set of equally probable forecasts (ensemble) and therefore also to quantify its space-time uncertainty. In this work, we limited ourselves to study the performance of deterministic forecasts, deliberately avoiding to discuss aspects related to space-time uncertainty, aspects that emerge from the discussion at the end of

Section 5. To allow the comparison also with the other methodologies we will use the deterministic forecast obtained averaging all the STEP members and we will indicate this estimate with ST hereafter.

3. Proposed Nowcast Technique

The nowcasting problem can be tackled with artificial intelligence techniques exploiting the recent results of neural network approaches in the fields of computer vision and natural language processing. As a matter of fact, radar imagery nowcasting can be thought as a supervised learning problem, where the parameters of a neural network are modified to minimize an objective function that depends on the differences between the estimated precipitation image and the real field. Among the various network architectures proposed in computer vision and, in some cases, also already applied to the problem of meteorological nowcasting, we opted for the PredNet model, developed in computer vision for the forecasting of future frames in a video sequence. A key insight of this and other deep neural network methods is that, in order to be able to predict how the visual scenery will change over time, an agent must have at least some implicit structural model of the object represented by the sequence of images and the possible transformations that these objects may undergo. In the case of precipitation nowcasting this translates into the need for the network to assimilate a representation of the movement of the clouds and the development of related precipitation. In the proposed architecture, this translates into the ability to interpret the cloud movements considering the advective part of the phenomena, and also predict variations in intensity, linked for example to the orography of the domain.

The PredNet architecture [23] continuously strives to predict the appearance of future video frames, using a deep and recursive convolutional network with connections from large scale to small scale (top-down) and vice versa. The approach exploits the experience of similar models based on recursive and convolutional networks for the prediction of successive video frames, but draws particular inspiration from the concept of "predictive coding" in neuroscience literature, which assumes that the brain continuously makes predictions by anticipating incoming sensory stimuli. The top-down connections of the PredNet network transmit these forecasts, gradually increasing the details, compared with real observations, to generate an error signal. This is then propagated back to coarser scales (bottom-up) across the network, ultimately leading to a forecast update.

The PredNet architecture used is shown schematically in Figure 1 and consists of a ladder structure with a descending branch that generates the prediction of the precipitation field image at progressively more detailed scales, and an ascending branch where the prediction is compared with the real data.

The generative block G_l receives as input a representation of the state coming from the upper block r_{l+1}, performs an upscaling (with factor 2) and concatenates the image thus obtained with the representation of the error e_l for that scale, calculated at the previous iteration and stored in memory. This data is passed to a recurrent Long Short-Term Memory cell with convolutional filters [19], which in turn retains a memory of its own state. The result of processing is the state at the l layer, r_l, where this image is further processed by a convolutional layer to get a representation of the precipitation h_l.

The discriminant block D_l receives as input the h_l precipitation forecast for the scale under examination, and the representation of the real precipitation field at the same scale a_l. From the comparison of the two images an error image is generated. This is stored in memory as the input of the generative block at the next iteration of the model, and is further processed by convolution and pooling, generating a representation of the a_{l+1} precipitation of lower detail, passed then to the upper discriminative block.

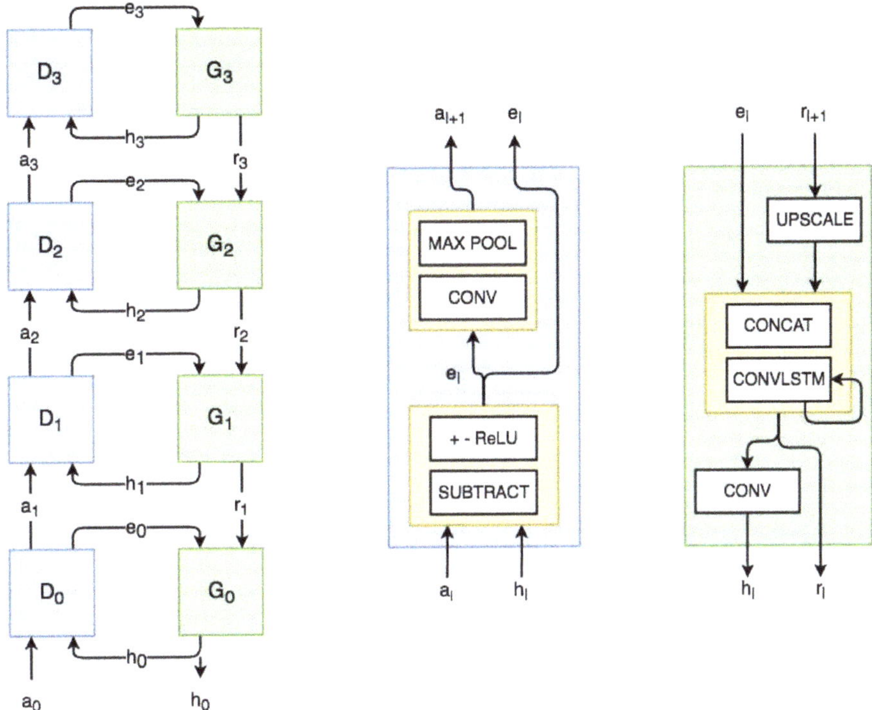

Figure 1. PredNet architecture: in the left the general architecture with the Discriminator blocks and the Generator blocks, in the right details of blocks.

Table 1 shows the tensor sizes a_l, h_l, r_l, e_l used in the model. The convolution filters have a kernel size of 3 with a padding of 1. The hidden state for the LSTM network has a size equal to the tensors r_l.

Table 1. PredNet tensors: size of the tensors for the different levels of the PredNet architecture expressed as (channels, width, height).

Level l	a_l	h_l	e_l	r_l
0	(1, 96, 96)	(1, 96, 96)	(2, 96, 96)	(1, 96, 96)
1	(16, 48, 48)	(16, 48, 48)	(32, 48, 48)	(16, 48, 48)
2	(32, 24, 24)	(32, 24, 24)	(64, 24, 24)	(32, 24, 24)
3	(64, 12, 12)	(64, 12, 12)	(128, 12, 12)	(64, 12, 12)

The model, as already said, provides more accurate results as new inputs are provided. The procedure presented here uses the images of the precipitation for the hour before the lead time, then the model is further iterated for the next hour's forecast times, replacing the real image of precipitation a_0 with the one that the model estimated h_0. With a time sampling of 5 min, 12 images a_0 for the first hour are provided and 24 images h_0 for the reconstruction of the first hour of precipitation and the estimates (without radar input) for the following hour are obtained by model iteration.

More details about the architecture can be found in Lotter, Kreiman and Cox [23]. This work is focused on forecasting the next frame in a video sequence, and it shows how the architecture is able to generate increasingly accurate forecasts as new frames are input, demonstrating to be able to refine forecasts and learn from mistakes made during the process. The training is performed with a Gradient Descent procedure in which the network parameters are changed to minimize e_l errors at the different levels of image representation with a loss function based on Mean Absolute Error. The results obtained in [23] show that in case of real video sequence the best results are achieved with the minimization of e_0 only, i.e., the MAE between the calculated and the actual image.

Based on those results here the loss function is released from the representation of the internal error in the network and the minimization of the Mean Squared Error between the image of the estimated and the actual precipitation is used.

Since the forecast is extended beyond the next frame, the loss function used in the training phase is realized as the Mean Squared Error of the 24 h_0 images generated by the network compared to the real a_0 precipitation images:

$$Loss = \sum_{t=0}^{24} MSE(a_0(t), h_0(t)) \tag{1}$$

where MSE is an operator that returns the Mean Squared Error between the actual radar image and the one estimated by the model; the sum is extended both to the 12 steps used for learning the current weather situation and to the 12 steps of the forecast, in order to train the network to "optimally" reconstruct both the radar images used as initial conditions and the predicted ones. Other experiments have also been carried out, using for example the MAE as loss function, obtaining, in the verification, an improvement limited to the specific metric (i.e., the MAE) but a more marked worsening for all the other metrics considered (i.e.,: RMSE, r and ETS). By using both MSE and MAE we verified that the training of a specific network for each forecast horizon did not produce an improvement in performance. Furthermore, tests were conducted using both the normalized rain intensity and a logarithmic transformation to feed NN. The best results were obtained using directly the color indexes of the image, and only after the predictive phase the rain rate transformation function described in Table 2. The weights of the neural network are optimized using a Gradient Descent procedure with Adam optimizer. Having data from 2013 to 2017 we used the images of the first 3 years for training, the images of 2016 for validation while the tests were conducted on the radar images of 2017. The neural network model and training procedures are implemented in python [32] using the PyTorch library [33] and are available in open source (https://github.com/lmssdd/RadarNowcast), the calculation was run on NVIDIA Tesla K80 GPU with 12GB of onboard RAM. The model will be referenced hereafter as NN.

4. Data

The data used in this application come from a meteorological radar located in the region between Osaka and Kyoto. These public domain data, already used in [21], have been collected from an online archive available on Yahoo (https://developer.yahoo.co.jp/webapi/map/openlocalplatform/v1/js/) and span five consecutive years from 2013 to 2017. Their spatial resolution is equal to 1 km^2, the temporal frequency is 5′, the domain covers an area of about 10^4 km^2 as shown in Figure 2 (https://www.jma.go.jp/en/highresorad/ and https://www.google.com/maps). The climate of the area is classified as hot-temperate, precipitation is intense (cumulative annual 1730 mm) and particularly heavy during the summer season as shown for 3 meteorological stations in Figure 3 (http://ds.data.jma.go.jp/gmd/tcc/tcc/products/climate/).

The dimension of the images obtained through the API is 1000 × 1000 pixels. After a crop necessary to remove text characters in the image, since the actual resolution of the radar is lower, a down-sampling of the image is performed, finally generating a 96 × 96 matrix. The data represent the precipitation intensity per pixel in a palette of 15 false colors. It is thus possible to map the colors as a progressive index of rain intensity with a range from 0 to 14 as reported in the Table 2.

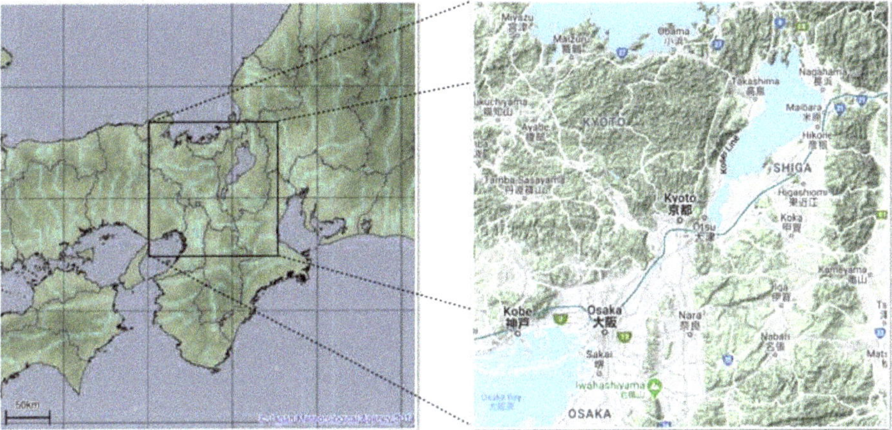

Figure 2. On the left the map of Japan in which is highlighted the area shown in detail on the right that represents the domain covered by the Japanese meteorological radar whose data is used in this work.

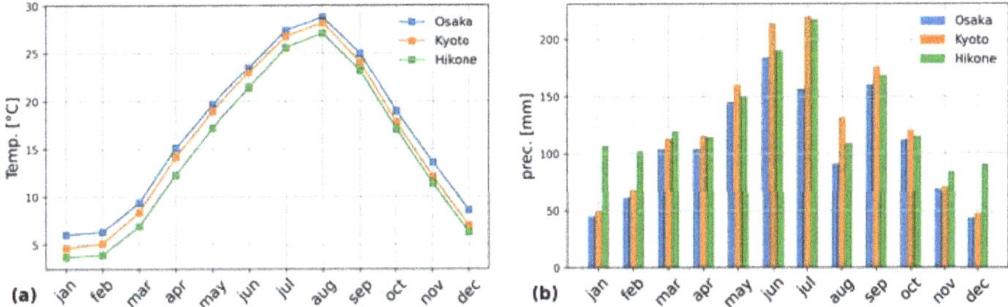

Figure 3. Climate data of monthly average temperature (**a**) and precipitation (**b**) for three meteorological stations in the domain.

Table 2. Table of correspondence between the colors of the radar image representation and the precipitation intensity.

Index	Color (RGB)	Rain (mm/h)
0	(255, 255, 255)	0
1	(236, 254, 252)	1
2	(196, 254, 252)	2
3	(156, 234, 252)	4
4	(156, 218, 252)	8
5	(180, 198, 252)	12
6	(180, 254, 156)	16
7	(180, 234, 156)	24
8	(164, 218, 156)	32
9	(252, 254, 156)	40
10	(252, 234, 156)	48
11	(252, 218, 156)	56
12	(252, 190, 196)	64
13	(252, 158, 156)	80
14	(228, 158, 164)	100

The array containing the color index is normalized to a maximum unitary value to be processed in the neural network, and it is mapped back to rain intensity values during the post-processing. The use of precipitation intensity, or its elaboration using a logarithmic transform, has proven to be less effective than direct processing of the color scale index. The data has also been filtered by eliminating all weather events where the radar images were incomplete (i.e., either totally or partially missing).

All other methods have been applied to the logarithm transform of the rain rate estimates and only on rainy events, defined as those with at least 5% of the domain with precipitation rate greater or equal to 1 mm/h. The comparison between the different methodologies is limited to 2017, to maximize the use of previous years for the neural network training and favour a direct comparison with the results of optical flow methodologies that, instead, do not require training.

In the operational phase, all methods applied need to process a different number of the most recent steps of the radar images. The Table 3 lists the number of time steps needed for processing and the number of forecast steps obtained by each procedure. The NN and SP methodologies need, as initial conditions, several steps equal to the number of forecast occurrences, while all others can work with only two time steps except ST that needs three steps as initial condition.

Table 3. Comparison of the number of images processed by the different methodologies to obtain a forecast of 12 steps, equivalent to 1 h.

Method	Past Steps	Forecast Steps
NN	12	12
ST	3	12
SP	12	12
DR	2	12
LK	2	12

5. Results

The performance of NN, ST, SP, DR, LK nowcasting techniques described in the Sections 2 and 3, were assessed for 2017. The 2013, 2014 and 2015 data were used for training, the data of year 2016 were used for validation of the NN method. The performance indicators were evaluated against the precipitation value, at forecast verification time as estimated by the radar itself. Therefore, although expressed in mm/h, they do not represent in any way an estimate of the predicted precipitation forecast error. As a matter of fact, they gauge the error of the radar measures once transformed into mm/h. The indicators used are: the root mean square error (RMSE), the mean absolute error (MAE), the Pearson correlation coefficient (r), and the equitable threat score (ETS) for precipitation exceeding 0, 1, 2, 4 mm/h respectively.

Before starting the discussion on the results, a preliminary consideration need to be made to understand both strenghts and limitations of nowcasts based on optical flow extrapolation techniques. All methods used (ST, SP, DR and LK) can fall into this category of course except the NN. Figure 4 shows the spatial autocorrelation of radar images as a function of the delay, in minutes, for 200 2017 "rainy" events, randomly chosen. As can be seen, the autocorrelation decays quickly but is appreciable (the orange line represents the inverse of the number of Nepero: $1/e$) up to a delay of about 60′. This can be taken as a limit of predictability for persistence, a particular case of optical flow with zero advection, and therefore it is expected the optical flow procedures to have non negligeable skills for up to one hour. For the same reason, 60′ is also the maximum delay that makes sense to use as input for any extrapolation procedure based on recent history. That is why both for the SP and NN procedure 12 images (at 5′ intervals) preceding the nowcast time were used as initial conditions.

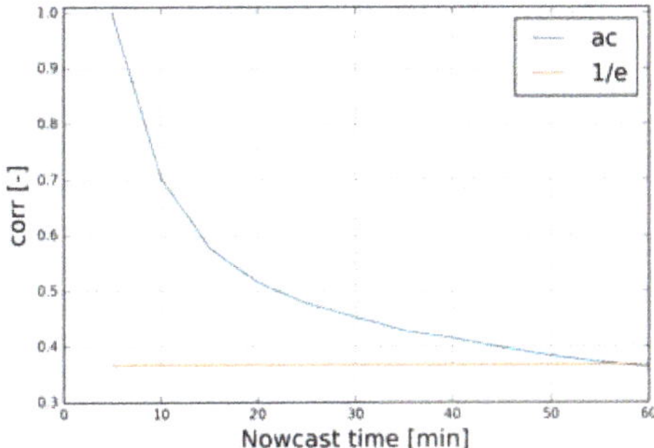

Figure 4. Autocorrelation of precipitation field as function of the delay expressed in minutes for a sample of 200 "rainy" events.

To clarify some aspects on which the discussion of the results will be based, in Figure 5 we show a sequence of images lasting one hour and related to the forecast of a precipitation event occurred on 8 January 2017. The first column shows the 12 frames that precede the forecast emission (train from minute −55 to 0), the second column (NN_train) shows the reconstruction of the precipitation field by the neural network. The network produces an estimate at each time step, starting from a zero field and gradually learning from the radar images provided. The third column (verif) shows instead the 12 images recorded by the radar and to be compared to in the following hour of forecast (minutes from 5 to 60). The following columns show, except the fourth (NN_flip) which we will discuss later, the nowcast for the procedures NN, ST, DR, SP and LK.

The domain of existence of the nowcast depends on the procedure used. The NN technique, being based on a neural network of generative type, returns a prediction for the whole domain while the SP and DR methods after the advective and extrapolative part try to "fill" the domain through interpolation, although generally this is never complete. Vice versa the ST and LK techniques, based only on the advection, have a field of existence just limited to the areas where the initial field has been "moved" while is undefined in the complementary areas. Note, in particular, in Figure 5 the frames for the forecast from 30′ onwards in which the area covered by the forecast is progressively reduced for all the forecasts except for NN. When comparing the performance of the different methods this fact is taken into account by evaluating the errors in the Domain of existence of each Single forecast (DS hereafter) and in the Domain resulting from their Intersection (DI henceforth). This problem is also highlighted by the extension of the domain covered by the radar data that is limited with respect to the typical advection speed of that climate zone and confirmed by the depicted event. This inconvenience could be avoided, or at least mitigated, considering a domain large enough to cover the central zone of greatest interest. To simplify the analysis of the results the discussion that follows is organized in subsections.

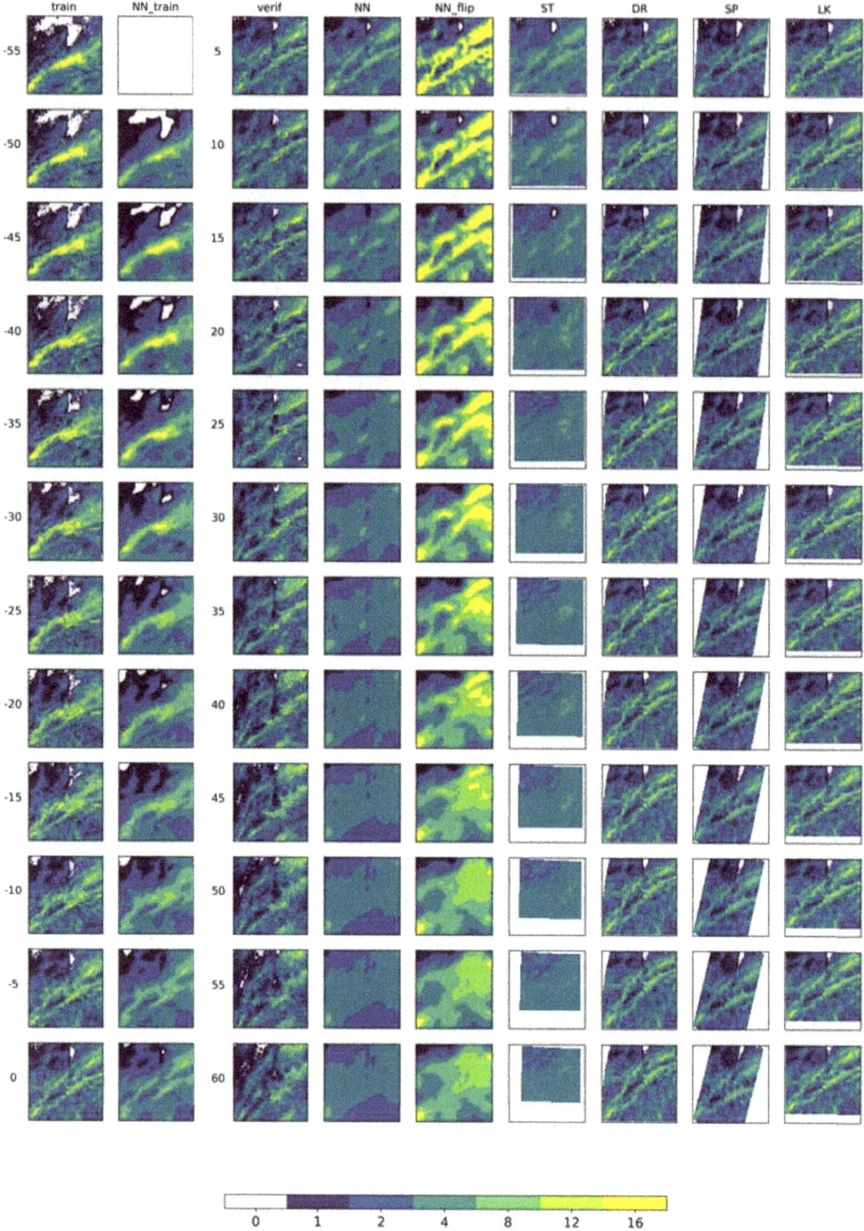

Figure 5. Frame sequence illustrating the mode of operation of the different nowcast techniques calculated for the one-hour forecast starting on January 8 at 16:25 UTM. See the text for a detailed description of the figure.

5.1. RMSE

The use of RMSE metrics is not completely appropriate, for a field as discontinuous as precipitation. Nonetheless it provides an interesting "bulk" measure to compare the performance of the various methods. The box (**a**) of Figure 6 shows the RMSE of the averaged forecast for the whole year 2017, depending on the forecast time from 1′ to 60′, for the procedures NN, ST, SP, DR and persistence, evaluated on the DS domains. It is clear that for all techniques there is an added value compared to persistence that lasts for the entire duration of the forecast. NN and ST have an RMSE significantly lower than all the other techniques and in particular ST prevails over NN only in the second half hour. However, since the mean RMSE has been evaluated and then averaged again over different domains for each technique, a direct comparison is not possible. For this reason, in Figure 6, on box (**b**), is always shown the average RMSE evaluated only on the domain DI. In this case the trend is confirmed but NN prevails over the other techniques for all the 60′ of duration of the forecast. The slight prevalence of ST over NN in the second half hour forecast can be related to ST's ability to incorporate the average effects of stochastic noise, whose influence increases with the forecasting time. On the other hand, NN has the clear advantage of providing a prediction for the entire domain covered by the radar. In addition this ST prevalence, as be shortly shown, disappears when the performance is evaluated using scores more suitable to assess the quality of the precipitation forecast, such as the MAE, r and, especially, the ETS.

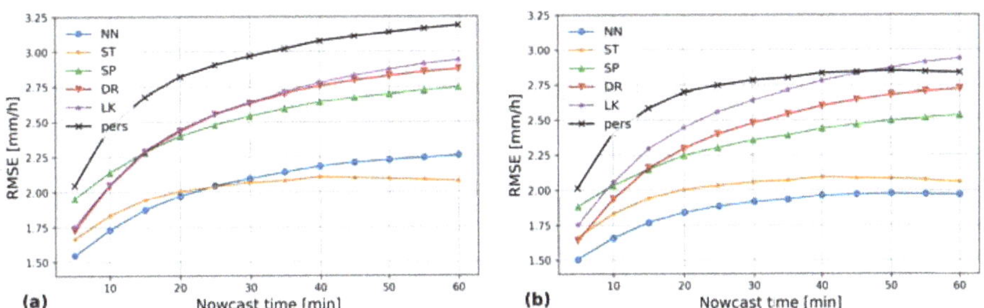

Figure 6. Average RMSE as function of the forecast time from 1′ to 60′, for all nowcast methods and persistence (2017). In box (**a**) the indicators are evaluated in the DS domain , in (**b**) the same indicators are calculated in DI.

5.2. MAE and Pearson Correlation Coefficient r

Figure 7 shows the values of the MAE and of the Pearson's correlation coefficient r as a function of the forecast time, from 1′ to 60′, for the nowcast of the 5 procedures and for persistence, for 2017. The superiority of NN over all other procedures is clear, both in terms of MAE and of r, even when these were calculated over the DS domain (boxes (**a**) and (**b**)). To verify the robustness of the results obtained, we repeated the calculation of the MAE and r, limiting the analysis to the DI domain. The results are shown in boxes (**c**) and (**d**) of Figure 5. In this case too the performance of the NN method proved to be better than those of each of the others.

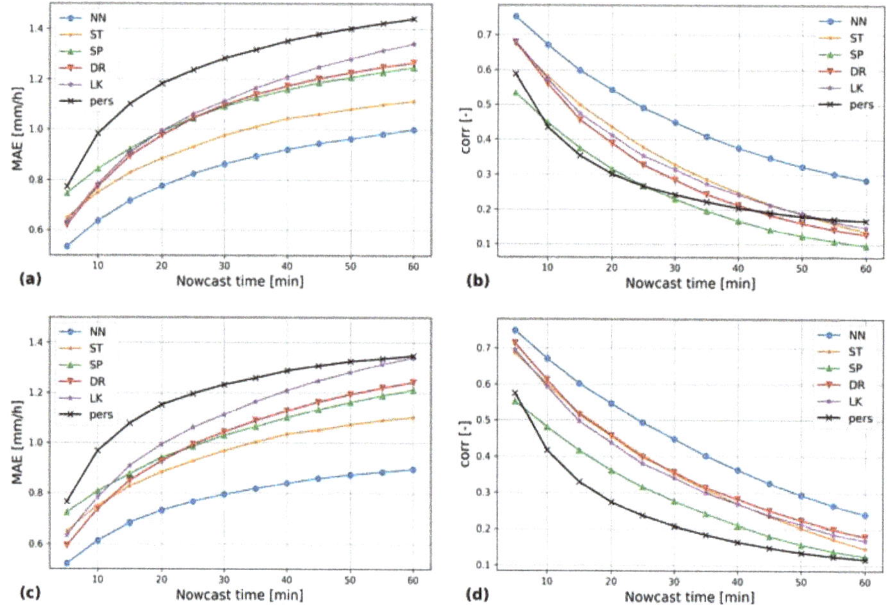

Figure 7. As in the previous figure for MAE and Pearson's correlation coefficient. First row of plots (boxes (**a**) and (**b**)) shows results evaluated in the domain DS, second row (boxes (**c**) and (**d**)) those in DI.

5.3. ETS

Figure 8 shows in boxes (**a**), (**b**), (**c**), (**d**) the ETS for the 0, 1, 2, 4 mm/h thresholds exceedance. While it is clear the advantage of the tested procedures with respect to persistence, the NN procedure stands out for the case of 0 mm/h (rain/no-rain) and 1 mm/h thresholds. For higher thresholds there is a slight prevalence of the LK method over all other procedures. However, the maximum values of ETS are significantly lower, as are the differences between the LK, NN and DR methods, which makes this lead not significant. The results reported in Figure 8 are obtained calculating the ETS in the DS domain (therefore different for each one), but the results obtained limiting the calculation to the common DI domain are basically identical and therefore not shown for brevity.

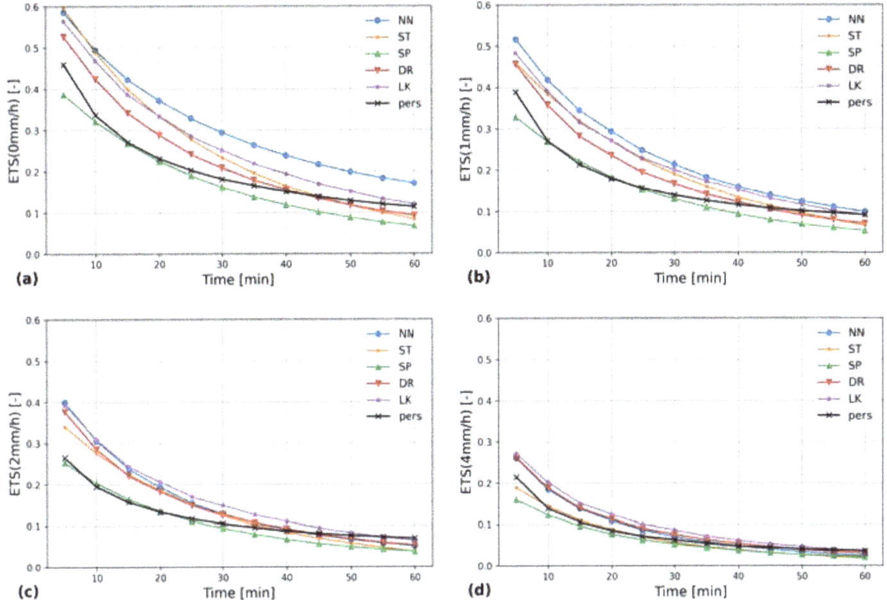

Figure 8. ETS, for 0, 1, 2, 4 mm/h thresholds shown on boxes (**a**), (**b**), (**c**), (**d**) respectively, as function of the forecast time from 1′ to 60′, for all nowcast methods and persistence, for 2017.

5.4. Skill Score

To have a measure of the accuracy of the nowcasting methods, for the entire duration of the forecast, we used two skill scores (SS) based on RMSE and MAE.

The accuracy is evaluated relatively to the persistence using the following espression:

$$SS(RMSE)_{meth} = 1 - \frac{RMSE_{meth}(t)}{RMSE_{pers}(t)} \tag{2}$$

where $meth = \{NN, ST, SP, DR, LK\}$ is the method for which the skill score is calculated. Similar formulas was used to calculate the SS(MAE). The values of the two SS obtained are summarized in Table 4. Looking at table, the plots and the discussion of the previous paragraph a clear superiority of the NN method is evident, despite the fact that both the RMSE and the MAE have been evaluated in the DS domain of each of the used methods. Once more, it is also worth to note that the only method able to produce a prediction on the entire radar domain is NN, while all others are limited to a portion of it, depending from the advection speed of the event and the forecast time. This is also the reason the SS(RMSE) for ST is slightly better than the one for NN (see discussion of Figure 6).

Table 4. RMSE- and MAE-based skill scores relative to the persistence for all nowcasting method used. Both RMSE and MAE are calculated within the domain DS of existence of each nowcast.

Method	SS (RMSE)	SS (MAE)
NN	0.29	0.32
ST	0.30	0.23
SP	0.13	0.14
DR	0.12	0.15
LK	0.11	0.14

5.5. NN vs. ST

The results clearly show that the NN nowcast procedure, based on generative neural network, has a significantly better performance than all the algorithm that, to the best of our knowledge, represent the current state of the art. At least in the chosen configuration: a domain of relatively small size, compared to the typical advection speed of the climate zone where the radar used operates. To investigate the origin of this performance difference, the input data for the nowcast were flipped with respect to the increasing direction of latitude and longitude and the nowcast repeated throughout the year 2017. All methods based on optical flow are invariant with respect to this transformation.

Not so for NN, as demonstrated by the MAE and r shown in Figure 9. In particular from a quantitative point of view, it is clear that the nowcast for the data flipped in latitude and longitude are completely wrong (see MAE for NN_flip on box (**a**)) while those for the correlation coefficient worsen considerably but remain in line with those of the other methods, in particular for forecast times over 30′ (see box (**b**)). Most likely this result is an indication of the neural network "learning" correctly, from the three years of training data, the climatic characteristics of the phenomena related to the particular domain. The behavior of NN applied to the flipped data, with respect to those on the natural domain, can be verified by the analysis of the test case shown in Figure 5. Comparing the NN frames with the corresponding NN_flip (see the fifth column of images), the similarity of the two predicted patterns is clear and might be related to the NN advective component. Regarding the amount of rain, NN_flip provides much more intense precipitation, and this can be related to the generative component of the network. This is consistent with the considerably higher values of MAE and, at the same time, with the r values similar to the other procedures.

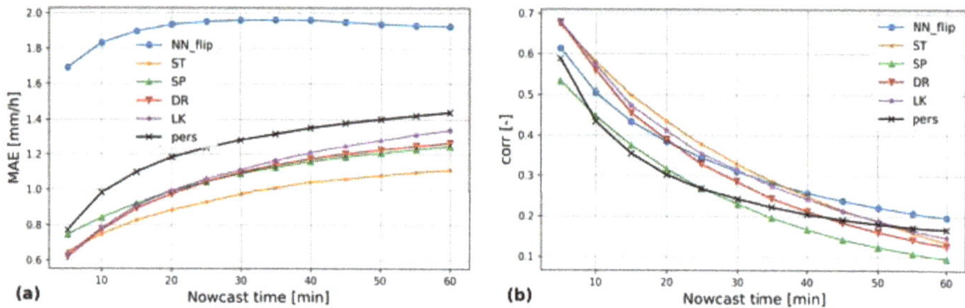

Figure 9. As in Figure 7. On box (**a**) the MAE and on box (**b**) the correlation coefficient of Pearson r as a function of the forecast time from 1 to 60 min, this time for the nowcast obtained by inverting the axes of longitude and latitude of the domain. The results for NN, the only ones that differ from those shown in Figure 7, are indicated with NN_flip.

5.6. Space-Time Behavior of the Nowcast

The improvement in performance achieved with NN brings in a reduction in the spatial detail of the forecast. This can be seen, from a qualitative standpoint, for the case shown in Figure 5, observing how the smoothing level of the solution NN increases with the lead time, more than for the other methods. Addressing the quantitative aspect, we calculated the spectral power density of the logarithmic transformation both for the measured and predicted fields In Figure 10 the results obtained by averaging the PSD on all the "rainy" events of 2017 for 5′, 30′ and 60′ lead times are shown on boxes (**a**), (**b**) and (**c**) respectively. NN starts to lose power already at a scale of about 16 km for the 5′ forecast. This loss of detail moves up to 32 km at 30′ lead time, and 48 km at 60′ lead time, in line with the results shown in [34] for the RainNet method.

This unintended feature of the NN forecast derives from the network optimization procedure which, while minimizing the loss function, devises "effective" smoothing the solution. The behavior, implicit at each 5′ forecast, is amplified by numerical diffusion as the forecast time progresses, because of the iterative nature of NN.

This is also reflected in NN's reduced ability to predict extreme values. As an example, in Figure 11, a time series of precipitation data, for a severe event (29 June to 2 July 2017), is shown for the point of the radar nearest to Kyoto. The radar data estimation of precipitation in mm/h are shown along with the corresponding NN prediction, at lead times of 5′, 30′, and 60′. The smoothing effect of the forecast could be seen also here, but now over time. For events longer than an hour, in this case the first and third day of the event, the procedure is able to reproduce the evolution of the precipitation field even at lead times of 60′. The central part, instead, is poorly predicted as it lasts less then the minimum needed learning time. The power spectra tells us that this is caused by the pixel scale not being resolved. However, for hydrological applications, if the basin has a surface as large as the scales resolved by the nowcast, this effect is minimized since all points within the basin contribute to the runoff.

The increasing space-time "smoothness" of the NN forecast with lead times, is related to the increasing uncertainty of the forecast, an aspect which is out of the scope of the present paper, and on which we are currently working on as an extension of the present research.

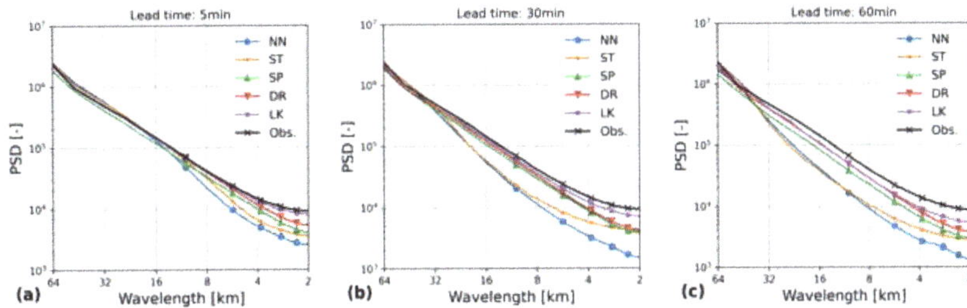

Figure 10. Power spectral density averaged over all the precipitation events of 2017, for times of forecast of 5′, 30′ and 60′ on boxes (**a**), (**b**) and (**c**) respectively.

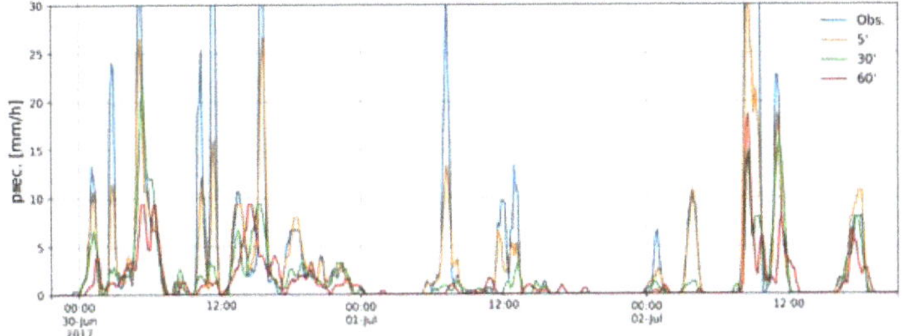

Figure 11. Time series of an event of intense precipitation occurred above Kyoto in the period from the 29 of June to the 2 of July 2017. Radar observations and corresponding NN forecasts at lead times of 5′, 30′ and 60′ are shown.

6. Conclusions

In this work, a comparison of state-of-the-art methods for the nowcast of precipitation intensity derived from weather radar images and based on optical flow, with a deep learning methodology based on PredNet architecture, trained with a specific loss function, is presented.

The detailed analysis of this extensive test case proves that the proposed algorithm, based on a generative neural network architecture, is far superior to any other method representing, to the best of our knowledge, the state of the art for this subject. It must be said, according to a precautionary principle, that these conclusions cannot be generalized, a priori, to the nowcast of the images of another type of radar or a similar one positioned in a different geographical region. This limit originates from the small size of the domain covered by the radar, compared to the typical advection velocity obtained through the application of OF techniques.On the other hand, there is no valid reason the procedure used here could not be used in other domains and/or other radar types, after an appropriate learning phase, to produce similar encouraging results.

It would be certainly interesting to further develop the proposed concepts and methodologies. In that respect a modification of the architecture in order to formally separate the advective component from the source/sink in the generative branch of the network will be applied in an upcoming work. This should facilitate a transfer learning process and build a general model, able to interpret and predict radar measures, that can then be specialized onto specific instruments and/or different meteo/climatic regions. We also intend to study the potential for a probabilistic nowcast, estimating the spatial and temporal uncertainty that can be obtained from neural network model, for instance through a quantile regression and the use of a pinball loss function [35].

Author Contributions: M.M. and L.M. jointly conceived and designed the methodologies, performed the analysis and wrote the paper. All authors have read and agreed to the published version of the manuscript.

Funding: This work was financed by the project "Tessuto Digitale Metropolitano" funded by the regional agency Sardegna Ricerche (POR FESR 2014-2020, Azione 1.2.2) and by Sardinian Regional Authorities.

Acknowledgments: The authors would like to thank Ryoma Sato of Kyoto University for sharing his scripts, necessary to download the dataset

Conflicts of Interest: The authors declare no conflict of interest.

References

1. Bartholmes, J.; Todini, E. Coupling meteorological and hydrological models for flood forecasting. *Hydrol. Earth Syst. Sci. Discuss.* **2005**, *9*, 333–346. [CrossRef]
2. Tran, Q.K.; Song, S.K. Computer Vision in Precipitation Nowcasting: Applying Image Quality Assessment Metrics for Training Deep Neural Networks. *Atmosphere* **2019**, *10*, 244. [CrossRef]
3. Heye, A.; Venkatesan, K.; Cain, J. Precipitation nowcasting: Leveraging deep recurrent convolutional neural networks. *Proc. Cray User Group (CUG)* **2017**, *2017*, 1–8.
4. Yu, W.; Nakakita, E.; Kim, S.; Yamaguchi, K. Improvement of rainfall and flood forecasts by blending ensemble NWP rainfall with radar prediction considering orographic rainfall. *J. Hydrol.* **2015**, *531*, 494–507. [CrossRef]
5. Verbunt, M.; Walser, A.; Gurtz, J.; Montani, A.; Schär, C. Probabilistic Flood Forecasting with a Limited-Area Ensemble Prediction System: Selected Case Studies. *J. Hydrometeorol.* **2007**, *8*, 897–909. [CrossRef]
6. Corral, C.; Sempere-Torres, D.; Revilla, M.; Berenguer, M. A semi-distributed hydrological model using rainfall estimated by radar; application to Mediterranean basins. *Phys. Chem. Earth B* **2000**, *25*, 1133–1136. [CrossRef]
7. Knöll, P.; Zirlewagen, J.; Scheytt, T. Using radar-based quantitative precipitation data with coupled soil- and groundwater balance models for stream flow simulation in a karst area. *J. Hydrol.* **2020**, *586*, 24884. [CrossRef]
8. Collier, C.G. On the propagation of uncertainty in weather radar estimates of rainfall through hydrological models. *Meteorol. Appl.* **2009**, *16*, 35–40. [CrossRef]
9. Nguyen, D.H.; Bae, D.H. Correcting mean areal precipitation forecasts to improve urban flooding predictions by using long short-term memory network. *J. Hydrol.* **2020**, *584*, 124710. [CrossRef]
10. Ayzel, G.; Heistermann, M.; Winterrath, T. Optical flow models as an open benchmark for radar-based precipitation nowcasting (rainymotion v0. 1). *Geosci. Model Dev.* **2019**, *12*, 1387–1402. [CrossRef]
11. Reyniers, M. *Quantitative Precipitation Forecasts Based on Radar Observations: Principles, Algorithms and Operational Systems*; Institut Royal Météorologique de Belgique: Brussel, Belgium, 2008.
12. Woo, W.C.; Wong, W.K. Operational application of optical flow techniques to radar-based rainfall nowcasting. *Atmosphere* **2017**, *8*, 48. [CrossRef]
13. Bechini, R.; Chandrasekar, V. An enhanced optical flow technique for radar nowcasting of precipitation and winds. *J. Atmos. Ocean. Technol.* **2017**, *34*, 2637–2658. [CrossRef]

14. Lucas, B.D.; Kanade, T. An iterative image registration technique with an application to stereo vision. In Proceedings of the Seventh International Joint Conference on Artificial Intelligence (IJCAI '81), Vancouver, BC, Canada, 24–28 August 1981; pp. 674–679.
15. Wong, M.; Wong, W.; Lai, E.S. From SWIRLS to RAPIDS: Nowcast applications development in Hong Kong. In Proceedings of the PWS Workshop on Warnings of Real-Time Hazards by Using Nowcasting Technology, Sydney, Australia, 9–13 October 2006.
16. Shi, J. Good features to track. In Proceedings of the IEEE Conference on Computer Vision and Pattern Recognition, Seattle, WA, USA, 21–23 June 1994; pp. 593–600.
17. Schneider, P.; Eberly, D.H. *Geometric Tools for Computer Graphics*; Elsevier Science: Cambridge, MA, USA, 2003.
18. Seed, A. A dynamic and spatial scaling approach to advection forecasting. *J. Appl. Meteorol.* **2003**, *42*, 381–388. [CrossRef]
19. Shi, X.; Chen, Z.; Wang, H.; Yeung, D.Y.; Wong, W.K.; Woo, W.C. Convolutional LSTM network: A machine learning approach for precipitation nowcasting. In *Advances in Neural Information Processing Systems*; Curran Associates, Inc.: Red Hook, NY, USA, 2015; pp. 802–810.
20. Shi, X.; Gao, Z.; Lausen, L.; Wang, H.; Yeung, D.Y.; Wong, W.K.; Woo, W.C. Deep learning for precipitation nowcasting: A benchmark and a new model. In *Advances in Neural Information Processing Systems*; Curran Associates, Inc.: Red Hook, NY, USA, 2017; pp. 5617–5627.
21. Sato, R.; Kashima, H.; Yamamoto, T. Short-Term Precipitation Prediction with Skip-Connected PredNet. In Proceedings of the 27th International Conference on Artificial Neural Networks (ICANN 2018), Rhodes, Greece, 4–7 October 2018; Springer: Cham, Switzerland, 2018; pp. 373–382.
22. Ayzel, G.; Heistermann, M.; Sorokin, A.; Nikitin, O.; Lukyanova, O. All convolutional neural networks for radar-based precipitation nowcasting. *Procedia Comput. Sci.* **2019**, *150*, 186–192. [CrossRef]
23. Lotter, W.; Kreiman, G.; Cox, D. Deep predictive coding networks for video prediction and unsupervised learning. *arXiv* **2016**, arXiv:1605.08104.
24. Pulkkinen, S.; Nerini, D.; Pérez Hortal, A.A.; Velasco-Forero, C.; Seed, A.; Germann, U.; Foresti, L. Pysteps: An open-source Python library for probabilistic precipitation nowcasting (v1. 0). *Geosci. Model Dev.* **2019**, *12*, 4185–4219. [CrossRef]
25. Bradski, G.; Kaehler, A. *Learning OpenCV: Computer vision with the OpenCV library*; O'Reilly Media, Inc.: Sebastopol, CA, USA, 2008.
26. Wolberg, G. *Digital Image Warping*; IEEE Computer Society Press: Los Alamitos, CA, USA, 1990; Volume 10662.
27. Bouguet, J.Y. Pyramidal implementation of the affine lucas kanade feature tracker description of the algorithm. *Intel Corp.* **2001**, *5*, 4.
28. Laroche, S.; Zawadzki, I. Retrievals of horizontal winds from single-Doppler clear-air data by methods of cross correlation and variational analysis. *J. Atmos. Ocean. Technol.* **1995**, *12*, 721–738. [CrossRef]
29. Germann, U.; Zawadzki, I. Scale-dependence of the predictability of precipitation from continental radar images. Part I: Description of the methodology. *Mon. Weather Rev.* **2002**, *130*, 2859–2873. [CrossRef]
30. Ruzanski, E.; Chandrasekar, V.; Wang, Y. The CASA nowcasting system. *J. Atmos. Ocean. Technol.* **2011**, *28*, 640–655. [CrossRef]
31. Pulkkinen, S.; Chandrasekar, V.; Harri, A.M. Fully spectral method for radar-based precipitation nowcasting. *IEEE J. Sel. Top. Appl. Earth Obs. Remote Sens.* **2019**, *12*, 1369–1382. [CrossRef]
32. Van Rossum, G.; Drake, F.L. *Python 3 Reference Manual*; CreateSpace: Scotts Valley, CA, USA, 2009.
33. Paszke, A.; Gross, S.; Massa, F.; Lerer, A.; Bradbury, J.; Chanan, G.; Killeen, T.; Lin, Z.; Gimelshein, N.; Antiga, L.; et al. PyTorch: An Imperative Style, High-Performance Deep Learning Library. In *Advances in Neural Information Processing Systems 32*; Wallach, H., Larochelle, H., Beygelzimer, A., d'Alché-Buc, F., Fox, E., Garnett, R., Eds.; Curran Associates, Inc.: Red Hook, NY, USA, 2019; pp. 8024–8035.
34. Ayzel, G.; Tobias, S.; Maik, H. RainNet v1.0: A convolutional neural network for radar-based precipitation nowcasting'. *Geosci. Model Dev.* **2020**, in review. [CrossRef]
35. Petersik, P.J.; Dijkstra, H.A. Probabilistic Forecasting of El Niño Using Neural Network Models. *Geophys. Res. Lett.* **2020**, *47*. [CrossRef]

Article

Are Combined Tourism Forecasts Better at Minimizing Forecasting Errors?

Ulrich Gunter [1,*], Irem Önder [2] and Egon Smeral [1]

1 Department of Tourism and Service Management, MODUL University Vienna, 1190 Vienna, Austria; egon.smeral@modul.ac.at

2 Department of Hospitality and Tourism Management, University of Massachusetts Amherst, Amherst, MA 01003, USA; ionder@isenberg.umass.edu

* Correspondence: ulrich.gunter@modul.ac.at; Tel.: +43-1-3203555-411

Received: 8 June 2020; Accepted: 25 June 2020; Published: 29 June 2020

Abstract: This study, which was contracted by the European Commission and is geared towards easy replicability by practitioners, compares the accuracy of individual and combined approaches to forecasting tourism demand for the total European Union. The evaluation of the forecasting accuracies was performed recursively (i.e., based on expanding estimation windows) for eight quarterly periods spanning two years in order to check the stability of the outcomes during a changing macroeconomic environment. The study sample includes Eurostat data from January 2005 until August 2017, and out of sample forecasts were calculated for the last two years for three and six months ahead. The analysis of the out-of-sample forecasts for arrivals and overnights showed that forecast combinations taking the historical forecasting performance of individual approaches such as Autoregressive Integrated Moving Average (ARIMA) models, REGARIMA models with different trend variables, and Error Trend Seasonal (ETS) models into account deliver the best results.

Keywords: Bates–Granger weights; uniform weights; (REG) ARIMA; ETS; Hodrick–Prescott trend; Google Trends indices

---

1. Introduction

The perishable nature of tourism products and services such as hotel overnights, airplane seats, or restaurant tables makes forecasting an important prerequisite for setting efficient strategies to ensure business success. The special characteristics of tourism products and services such as perishability, intangibility, and consumption at the point of service delivery, external factors such as natural and man-made disasters, as well as unsteadiness of human nature make forecasting an important issue for international government bodies, national governments, academics, and practitioners alike.

In the past decades, many studies have dealt with the challenge of improving tourism demand forecasting accuracy, yet all these research efforts have led merely to the conclusion that no single forecasting method outperforms all others in all situations [1]. Furthermore, discussions of complexity have become increasingly relevant in the academic literature aiming to improve forecasting accuracies. Green and Armstrong [2] note that the trend to develop increasingly complex approaches has a long history, yet is at odds with scientific principles that advocate simplicity. An alternative way to use complex techniques to improve forecasting accuracy would be to combine the forecasts of individual forecasting models with the help of various combination techniques, as it has been shown that combined methods minimize the risk of extreme inaccuracy by "averaging out" the weaknesses of single models [3]. Forecast combinations are also capable of introducing adjustments and additional information balancing out measurement errors, which could negatively affect forecasting power [4].

Despite the many tourism studies carried out to date, the application of forecast combination techniques as a tool to create complexity remains rare. In contrast to other disciplines, research into combination methodologies for tourism demand forecasting has a short history, which started only in the early 1980s. Research into combining forecast methodologies was stimulated significantly earlier in various other economic and business fields by the seminal work of Bates and Granger [5]. They examined the performance of combining two sets of forecasts of airline passenger data, whereby the weights of the individual forecasts were calculated based on the historic predictive performance of each individual approach, and found that the combined forecasts showed lower errors than the individual forecasts. Only 20 years after the 1969 Bates and Granger paper, Clemen [6] summarized the intensive work that had been done around the topic of forecast combinations in the interim, and delivered an encompassing literature survey about these activities.

The manageable number of studies about forecast combinations in tourism have, in general, delivered the outcome that the combined forecast outperformed the forecasts generated by individual models [3,7–13]. Our work compares forecasting accuracies for eight quarterly report periods spanning the period from March 2015 until August 2017 in order to check the stability of the results during a changing macroeconomic environment. In other words, the forecast evaluation exercise is not only carried out once for different forecasting horizons but eight times recursively (i.e., based on expanding estimation windows), which corresponds to a "natural" practitioner's situation.

To meet these different challenges, we developed forecasting models for arrivals and overnights and used Eurostat data according to our commissioned study for the European Union as a whole. These models were tested across eight different quarterly periods for their stability and accuracy. In addition, we calculated combined forecasts based on the forecasts produced by the various individual forecasting models, and assessed the accuracies of these combined forecasts. Therefore, the objective of this study was to analyze whether combined forecasts of selected models are able to outperform the forecasts generated by the individual models in terms of forecasting accuracy.

Forecasting for the European Union as a whole for the indicated period and predicting three and six months ahead, as well as using both single forecasting models and forecast combination techniques that are easily replicable by practitioners were requirements by the European Commission, which contracted the present study [14]. Tourism plays a major economic role in the European Union: 13.6 million people (9.5% of all employees in the non-financial sector) were employed by 2.4 million tourism businesses (around 10% of all non-financial businesses) in 2016 [15]. In the same year, the tourism industry accounted for 3.9% of turnover and 5.8% of value added in the non-financial sector.

The remainder of this paper is as follows: After the literature review (Section 2), we discuss the advantages of forecast combinations and the most-used combination methods (Section 3). In the methodological section (Section 4), we present the chosen modeling approaches, the accuracy measures and their application, as well as the data used. The subsequent section provides and discusses the forecasting performance of the different approaches (Section 5). Conclusions and recommendations are the final section of the paper (Section 6).

2. Related Literature

The prediction competition for time series forecasting already has a history of about 50 years. According to Hyndman [16], the earliest scientific study of time series forecasting accuracy—the Nottingham study—was done by David Reid [17]. Paul Newbold and Clive Granger took the next step by conducting a study of forecasting accuracy involving 106 time series [16,18]: one important result of their study was that forecast combination improves forecasting accuracy. Comparing forecasts became fashionable and, over the years, "forecasting competition" has become an important term in the forecasting literature.

The first big forecasting competition took place in 1982 and was organized by Spyros Makridakis and Michele Hibon. For this competition—known in the forecasting literature as the "M(akridakis) competition" —anyone could submit forecasts related to 1001 time series taken from demography, industry, and economics [16,19]. The following M-2 competition was organized in collaboration with four companies, this time using only 29 time series and with the main purpose of simulating real-world forecasting. A peculiarity of this M-2 competition was that forecasters were allowed to use personal judgements, to ask questions about the data, and to revise their previous forecasts for the next forecast [20]. The succeeding M-3 competition had the objectives of replicating and extending the features of the previous competitions, with more methods, more researchers, and more (i.e., 3003) time series [21]. The most recent competition that had already been completed while this study was written, M-4, ran from January to May 2018, used 100,000 time series, and considered all major forecasting methods, including those based on Artificial Intelligence, as well as traditional statistical ones [22]. Informing the major objective of our paper, a stable result across all the M competitions was that combined approaches, on the average, outperformed individual forecasts [17–22].

Other competitions have also been organized in parallel to the M competitions. Mathematicians and physicists interested in forecasting ran their own competition at the Santa Fe Institute, beginning in January 1992 [23]. Other examples for forecasting competitions are the application of neural networks [24] or the global energy forecasting competitions [25,26].

In tourism, research into combination approaches and their efficacy started significantly later than in other disciplines. One of the first studies about forecast combinations in tourism was that by Fritz et al. [10] about the combination of time series and econometric forecasts. Their paper presents parsimonious methods of improving forecasting accuracy by combining various forecasting techniques. The Box-Jenkins stochastic time-series method was combined with a traditional econometric technique to forecast airline visitors to the State of Florida. Some years later, Calantone et al. [8] confirmed the results of Fritz et al. [10] and showed, also for the State of Florida, that forecasts of tourist arrivals obtained by forecast combination were more accurate than forecasts based on individual approaches. Shen et al. [3] tested the accuracy of forecast combinations compared to the forecasting results of seven different techniques over different forecasting horizons and demonstrated that combinations were superior to the best of the individual forecasts. Song et al. [27] shed further light on these results by showing that combined forecasts may be more beneficial for long-term forecasting. Shen et al. [28] compared six different combination methods and found that those that consider the historical performance of individual forecasts perform better than simple uniform average methods. In contrast, Gasmi [11] demonstrated that from three combination techniques, the Granger-Ramanathan regression method [29] delivered superior results in comparison to the simple uniform average technique and the Bates–Granger variance-covariance technique [5].

Andrawis et al. [7] suggested combining forecasts based on diverse information using different time aggregations (e.g., monthly and annual data). In comparing several forecast combination techniques, they show that the approach using forecasts based on time series with diverse time aggregations outperformed the combined individual forecasts based on time series with the same time structure. For improving tourism forecasts, Cang [30] proposed a non-linear combination method using multilayer perception neural networks, which can map the non-linear relationship between inputs and outputs.

On the other hand, a minority of studies has shown that combined forecasts do not always outperform the best individual forecasts, but are almost certain to outperform the worst individual forecasts [27]. Furthermore, Song et al. [27] stated that combined methods outperform the best single forecast in fewer than 50% of cases on average. A few years later, Song et al. [31] similarly stated that according to their results, forecast combination only improved forecasting performance in the tourism context in just over 50% of all cases compared with the most accurate single prediction. In a similar vein, Gunter and Önder [12] found that combined forecasts based on Bates–Granger weights, on multiple forecast encompassing tests, as well as on a combination of the two approaches [32]. The authors applied the aforementioned forecast combination techniques to Google Analytics indicators used as leading indicators for forecasting tourist arrivals to Vienna. However, these quite complex techniques only performed well for longer forecasting horizons.

To the best of the authors' knowledge, only one true forecasting competition focused on tourism time series data has been held to date [33]. The data set included 366 monthly, 427 quarterly, and 518 annual time series, all supplied by either tourism bodies or academics who had used them in previous tourism forecasting studies. The forecasting methods implemented in the competition were univariate and multivariate time series approaches, and econometric models. Surprisingly, however, this competition did not evaluate the accuracies of combined forecasts compared to individual forecasts.

3. Advantages of Forecast Combination

Why do combined forecasts perform better than individual forecasts in many contexts? Bates and Granger [5] stated that the simple portfolio diversification argument justifies the idea of combining forecasts. Forecast combinations offer diversification gains that make it efficient to combine individual forecasts rather than taking forecasts from just one single model. The information set underlying the individual forecasts is often unobservable for the forecast user: potentially because it comprises private information. Differences between the subjective judgements of various forecasters could therefore reflect differences in their respective information sets. In this situation, it is not possible to pool the underlying information set and construct a superior model that captures each of the underlying forecasting models. On the other hand, the higher the degree of overlap in the information set used to produce the underlying forecasts, the less useful a combination of forecasts is likely to be [34]. Furthermore, when forecast users have access to the full information set used to construct the individual forecasts, combinations are sub-optimal and it might be better to recommend finding a superior single model [35,36].

A second reason for using forecast combinations is that individual forecasts are differently influenced by structural breaks caused, for example, by institutional change or technological developments. Some models may adapt fast and only be affected by the structural break for a short time, while other models have parameters with slower adaption speeds. Since it is typically difficult to detect structural breaks in real time, it is plausible that, on average–i.e., across periods with varying degrees of stability–combinations of forecasts from models with different degrees of adaptability will outperform forecasts from individual models [37]. Similarly, Stock and Watson [38] report that in cases of structural breaks, the performance of combined forecasts tends to be far more stable than that of individual forecasts.

Third, forecast combination could be viewed in the sense that additional forecasts act like intercept corrections (ICs) relative to a baseline forecast. ICs can improve forecasting not only if there are structural breaks, but also if there are deterministic misspecifications [39].

Fourth, pooling of forecasts can also be understood as a shrinkage estimation. According to this approach, the unknown future value is viewed as a "meta-parameter" of which all the individual forecasts are estimates [39]. In these cases, averaging may improve the estimates.

Often, we also measure the wrong things. Demand data are rarely, if ever, available: thus, instead of measuring demand, we measure supply data (e.g., in periods with over-utilization of production capacities). However, it is obvious that such proxies of apparent demand introduce systematic biases in measuring real demand and therefore increase forecasting errors [4]. Averaging forecasts, in turn, would balance out these potential errors. Similar conclusions can be drawn for measurement errors and unknown misspecifications.

Moreover, statistical models assume that patterns and relationships remain constant. This is not always given: especially in the real world, where events and actions or fashions bring systematic changes and therefore introduce non-random errors in forecasting. Combining forecasts would help to increase their accuracy. Combining is also expected to be useful when experts are uncertain of which method to choose. This may be because we encounter novel situations (e.g., Brexit, the COVID-19 pandemic, stock market crashes, etc.) or have to make forecasts for a longer time horizon.

A further argument for combining forecasts is that the underlying forecasts may be based on different loss functions [40]: let us assume there are two forecasters, both have the same information set for forecasting a specific variable; however, forecaster #1 dislikes large negative forecasting errors, while forecaster #2 dislikes large positive forecasting errors. As a consequence, forecaster #1 will under-predict, while forecaster #2 will over-predict. If the bias is not constant over time, a forecast user with a rather symmetric loss function would find a combination of these two forecasts better than following the individual ones.

While forecast combination has advantages, there are also several arguments against it. Following the literature, forecast combinations are at a disadvantage over a single forecasting models because they produce parameter estimation errors in cases where the weights to combine the different forecasts need to be estimated [40]. This important consideration of avoiding errors in estimating the weights for the forecast combination has led simple uniform weighing methods to dominate more complex combination methods in mainstream scientific practice. The important advantage is that the weights are known and therefore do not have to be estimated: this plays a role if there is little evidence on the performance of the individual forecasts or if the parameters of forecasting model are time-varying. Furthermore, in many situations, a simple uniform average of forecasts will result in a significant reduction in variance and bias through averaging out the individual biases [40,41]. In the literature, the most used simple combination approaches are the Simple Average Combination (SAC) method and the VAriance-COvariance (VACO) method [3,5].

The SAC method assigns equal weights to each of the individual forecasts instead of using any optimal weights to minimize the variance of the combined forecasts. Although forecast combinations with equal weights could be biased, they might contribute to the reduction of the forecasting error as these weights are not influenced by other errors accruing from the estimation of optimal weights [3,42]. According to Palm and Zellner [41], the SAC method has the following advantages. When there is little evidence on the performance of the individual forecasts, it is an important advantage that the weights are known and therefore, no estimation is necessary. Furthermore, in many situations, the application of the SAC method contributes to the reduction of the variance and the bias through averaging out individual biases. Another advantage of the SAC method is its avoidance of sampling errors and model uncertainty in estimating optimal weights.

According to the VACO method, past errors of each individual forecast are used to determine the weights in forming the combined forecasts [5,6]. Bates and Granger [5] suggest assigning higher weights to good forecasts (low errors) and lower weights to poor ones (high errors).

4. Forecasting Tourism Demand for the European Union

The objective of the commissioned study based on the research period 2005–2017 was to find the model best at forecasting arrivals and overnights for the total European Union in the short term. The competing models should have a low degree of complexity, as the "winning" model was to be applied by the European Commission and Eurostat to an actual database in order to look ahead into the near future and to mitigate the lack of tourism data reporting [14]. This study was designed for the European Commission in order to help them forecast tourism demand for the European Union as a whole. Thus, similar models were used to test the forecasting accuracy at each time period.

In doing so, we analyze if the combined forecasting approaches according to the SAC and VACO methods based on the outcome of Autoregressive Integrated Moving Average (ARIMA) models, REGARIMA models with different trend variables, and Error Trend Seasonal (ETS) models (a state-space framework comprising traditional exponential smoothing models) outperform the single models, in general and over all periods, or if specific individual models show superiority [4,43–45]. Similar to the choice of the forecast combination techniques, the single forecasting models were also chosen based on the criterion of easy replicability by practitioners.

The database for the model estimations comprised the monthly arrivals and overnights statistics from Eurostat for the total of the EU-27 (unfortunately, no data were available for Ireland) from January 2005 until August 2017. In the estimations and forecasts, we separated between non-residents and residents (see also Figures 1 and 2). For calculating the forecasting accuracy, we performed out-of-sample forecasting for three and six months ahead. For the computations, we used EViews 9.5, and EViews 10 for the final quarterly report.

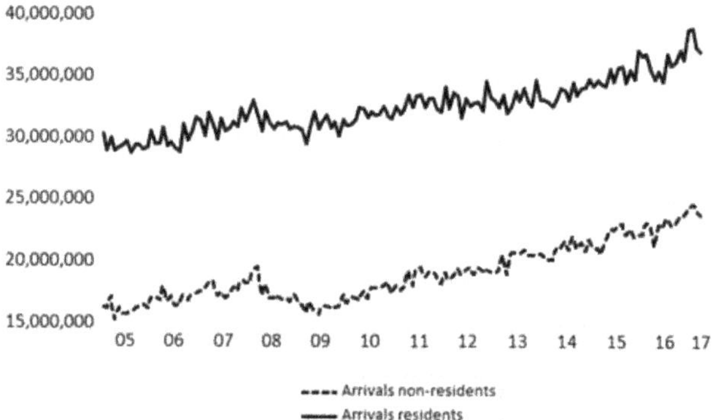

Figure 1. Arrivals of residents (solid line) and non-residents (dashed line) for the EU-27 (seasonally adjusted by the multiplicative ratio to moving average method). Source: Eurostat and own illustration.

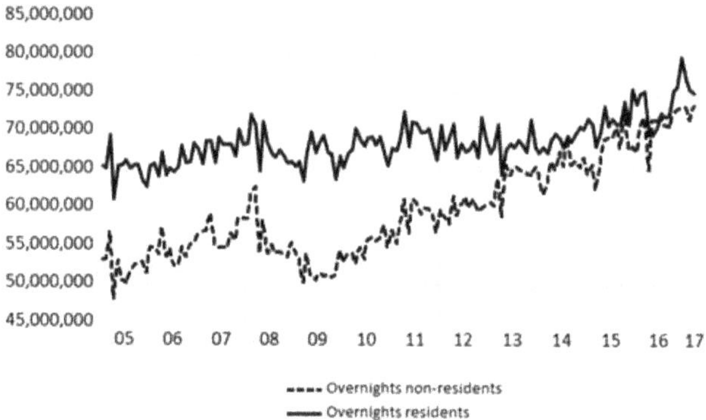

Figure 2. Overnights of residents (solid line) and non-residents (dashed line) for the EU-27 (seasonally adjusted by the multiplicative ratio to moving average method). Source: Eurostat and illustration.

4.1. An Outline of the Models Used

For solving the forecasting problem, we used four different approaches:

1. Autoregressive Integrated Moving Average (ARIMA) models,
2. REGARIMA models with different trend variables,
3. Error Trend Seasonal (ETS) models, and
4. Combined forecasts (with Bates–Granger weights (VACO) and uniform weights (SAC)) based on the forecasts produced by the different single forecasting models mentioned above.

4.2. ARIMA and REGARIMA Models

A general ARIMA (p, d, q) model for a first and seasonally differenced forecast variable $\nabla\nabla^s y_t$ (i.e., $d = 1$) in period t $(t = 1, \ldots, T)$ reads as follows:

$$\varphi(L)\nabla\nabla^s y_t = a + \vartheta(L)e_t \quad (1)$$

where $\varphi(L)$ and $\vartheta(L)$ in Equation (1) denote lag polynomials of finite orders p and q, while a denotes the intercept and e_t denotes the random error term. In this study, y_t corresponds either to overnights or to arrivals for the total EU-27.

A general REGARIMA (p, d, q) model for a first and seasonally differenced forecast variable $\nabla\nabla^s y_t$ (i.e., $d = 1$) with a contemporaneous exogenous variable $trend_t$, in turn, reads as follows:

$$\varphi(L)\nabla\nabla^s y_t = a + \vartheta(L)e_t + trend_t \quad (2)$$

where the notation in Equation (2) corresponds to that of Equation (1). In this study, $trend_t$ corresponds either to a Hodrick–Prescott trend of the arrivals and the overnights or to various Google Trends indices. It should be noted that the forecast variable is only first or second-differenced in the REGARIMA models with Google Trends indices, but not seasonally differenced.

4.3. Employed ARIMA Models

For all variables used in the ARIMA models, we applied seasonal differencing to remove any deterministic or stochastic seasonal patterns. Because of the existence of non-seasonal unit roots, we additionally took first differences to achieve difference-stationary processes.

For the three and six month forecasting periods and to achieve the best model fit, we employed an ARIMA (2, 1, 0) approach to model the overnights of residents and non-residents. To model the arrivals of non-residents, we applied the ARIMA (2, 1, 1) model, while we chose the ARIMA (2, 1, 0) model for the residents. All the estimated coefficients were statistically significant. The estimated equations also had excellent results in terms of out-of-sample forecasting accuracy (see as an example the forecasting accuracies for arrivals and overnights for the study period December 2017 in Tables A1–A8 in Appendix A).

4.4. Employed REGARIMA Models with Hodrick–Prescott Trends

In order to manage the short-term forecast problem, a quasi-causal model was constructed to explain arrivals and overnights. This model is based on a REGARIMA approach, which uses the flexible trend of the overnights/arrivals being explained through the model as its contemporaneous exogenous variable. The flexible trend was identified by the Hodrick–Prescott (HP) filter method and indicates important exogenous aggregated information in the model [43,46]. Correcting the forecast based on the flexible trend with AR and MA errors optimized the results.

In doing so, the overnights/arrivals values were transformed into absolute previous year differences of the moving 12-month averages of the log-transformed original values. 12-month averages were used to adjust for seasonal fluctuations, calendar effects, and special events. The explanatory variables are absolute previous year differences of the flexible overnights/arrivals trends identified by the HP filter method [43,46]. As the HP filter is based on the moving 12-month average of the log-transformed original values, these variables are easy to extrapolate by using exponential smoothing methods for forecasting purposes [4].

For the three and six month forecasting periods, we employed REGARIMA approaches and corrected the processes with AR (1) errors for modeling non-residents' and residents' overnights. To model the arrivals of non-residents, we applied approaches corrected with MA (1) errors, while for the residents we corrected with AR (1) errors. All the estimated coefficients were statistically significant. The estimated equations also had excellent results in terms of out-of-sample forecasting accuracy.

4.5. Employed REGARIMA Models with GoogleTrends

Based on the REGARIMA models outlined in the previous section, we also used model variants with Google Trends indices instead of HP trend variables. In order to achieve stationarity, we took the first or second difference of the variables (no seasonal differencing was performed). First differences were taken of arrivals, while the Google Trends indices were used as they came. Second differences were taken of overnights, while the Google Trends indices were employed in first differences.

For the three and six month forecasting periods, we employed the aforementioned REGARIMA approaches. Various correction processes were necessary for both the arrivals and overnights data.

To model the arrivals of the non-residents, we applied approaches corrected with AR (2) and MA (1) errors, while for the residents, we corrected with AR (2) errors. For modeling overnights, we applied AR (2) errors throughout.

Next, we explain what Google Trends are, where we can find them, and how we can use them. Google provides search data at an aggregated level (as an index) on its Google Trends page (http://trends.google.com/trends/), where users can identify the topics trending in search results or investigate a specific search term to learn about its popularity in different parts of the world. These data are open and free of charge to Google account holders, and can be downloaded in common spreadsheet formats to be used for analytical purposes, including forecasting.

In order to determine which web search term was more useful, we collected and developed four types of Google Trends variables: (i) a Google Trends index with country name (EU_trends), (ii), a Google Trends index with country names and flights (EU_flights), (iii) a Google Trends index with country names and hotels (EU_hotels), and iv) a Google Trends index with country names, flights, and hotels (EU_travel). There is not a consensus among researchers about how to choose the keywords for analysis. One method was directly choosing the keywords by subjective assessment of a set of text or data [47] and in this study we used the keywords that are related to travel planning (i.e., flights, hotels) under the travel category of Google Trends.

To calculate the EU_trends variable, the name of each of the 27 European Union countries was used as the search term to retrieve the respective monthly search indices from the Google Trends website. The data were retrieved in monthly intervals between January 2016 and December 2017 and compiled in a single Excel file. Then, to generate a regional EU_trends variable, the average of the index values across the 27 countries was calculated for each month separately. The EU_flights and EU_hotels variables were also calculated in a similar way: the search terms being each country's name followed by "flights" or "hotels", respectively. As a last variable using Google Trends indices, we generated an EU_travel variable by summing all the search indices used for the previously calculated monthly "EU" indices, such that EU_trends + EU_hotels + EU_flights = EU_travel.

In most cases, the estimated models had statistically significant parameters and satisfactory forecasting accuracy results (for an example see Tables A1–A8 in Appendix A). The reason why insignificant variables were retained was so the models could be tested using the same variables for the whole duration of the study period. A second reason was that, although some variables were indeed statistically insignificant in specific periods but not over the whole duration time of the study, this does not imply that these variables are unimportant for forecasting. In doing so, we follow a recent statement of the American Statistical Association (ASA) pointing out very clearly that the widespread use of statistical significance—to be understood as a 5% p-value threshold—as a justification for scientific findings leads to a biased perception of the scientific process [48]. The ASA statement has been an impulse to the scientific community to move further toward a world beyond $p < 0.05$ and a signal to recognize that statistical inference is not equivalent to scientific inference [49,50].

4.6. Error Trend Seasonal (ETS) Models

The Error Trend Seasonal (ETS) model class was developed by Hyndman et al. [45,51] and encompasses various well-known exponential smoothing methods (e.g., single exponential smoothing, double exponential smoothing, additive and multiplicative seasonal Holt-Winters) within a theoretically founded state-space framework which is estimated recursively by employing maximum-likelihood methods. The ETS framework consists of a signal equation for the forecast variable and a number of state equations for the three components that cannot be directly observed: level, trend, and seasonal. Since the ETS framework can automatically detect both trend and seasonal patterns and apply the most suitable model, further data transformation was not necessary.

Generally speaking, an ETS($\cdot,\cdot,\cdot$) model is represented by one of the following configurations of the error, trend, and seasonal components of the forecast variable, i.e., total EU-27 overnights or arrivals in this study [52]:

$$E(Error) \in \{A, M\},\ T(Trend) \in \{N, A, A_d, M, M_d\},\ S(Seasonal) \in \{N, A, M\} \tag{3}$$

where A in Equation (3) corresponds to additive, A_d to additive damped, M to multiplicative, M_d to multiplicative damped, and N to none.

This makes a total of 30 possible ETS specifications. From these, the most suitable specifications were automatically selected by employing information criteria such as the Akaike information criterion (AIC) and the Schwarz or Bayesian information criterion (BIC). Information criteria such as AIC and BIC are means for model selection and offer a relative estimate of the information lost in terms of the log likelihood function when a given model, e.g., a particular ETS specification relative to all other ETS specifications, is used to represent the process that generates the data [44]. Here, AIC and BIC have been calculated for all 30 possible ETS specifications, and the specifications characterized by the minimum AIC value and BIC values were then used for estimation and forecasting.

As an example of an ETS specification for three months ahead forecasting of overnights of non-residents in the EU-27 as selected by BIC, please see Table A5 in Appendix A. The one signal and two state equations of the selected ETS (M, N, A) specification read as follows [52]:

$$y_t = (l_{t-1} + s_{t-m})(1 + e_t) \tag{4}$$

$$l_t = l_{t-1} + \alpha(l_{t-1} + s_{t-m})e_t \tag{5}$$

$$s_t = s_{t-m} + \gamma(l_{t-1} + s_{t-m})e_t \tag{6}$$

where Equation (4) corresponds to the signal equation for the forecast variable y_t, Equation (5) to the state equation for the unobservable level component l_t, and Equation (6) to the state equation for the unobservable seasonal component s_t. α and γ denote the two smoothing constants, while the remaining notation in Equations (4) to (6) corresponds to that of Equations (1) and (2). It should be noted that the ETS(M, N, A) specification does not contain a state equation for the unobservable trend component since it is not present.

4.7. Forecast Combinations

Apart from the individual forecasting models, also the merits of two forecast combination techniques are evaluated. Bates and Granger [5] indicate that combination forecasts can yield lower forecasting errors, a finding which was later confirmed by Clemen [6]. To this aim, the forecasts produced by the 64 models, separately for forecast horizons three months and six months ahead, were all combined (or averaged) based on two methods that are common in the literature: simple uniform weights (unweighted average, or SAC) and so called Bates–Granger weights (weighted average, or VACO). On the other hand, there are more complex combination approaches available, with the given disadvantage that additional errors through parameter estimations might flaw the results. To avoid these additional errors sources, we used only combination methods based on calculated weights in this study.

More formally, a simple uniformly combined forecast F_h^U of overnights/arrivals for non-residents/residents for a forecast horizon h ($h = 3, 6$) is calculated as follows:

$$F_h^U = \sum_{m=1}^{M} \frac{1}{M} F_h^m \tag{7}$$

where F_h^m in Equation (7) is a forecast value produced by one of the single m ($m = 1, \ldots, M$) competing single forecasting models.

The Bates–Granger weight of an individual forecasting model is calculated as the inverse of the mean square error of that forecasting model relative to the sum of the inverses of the mean square errors of all forecasting models. Hence, a better individual forecasting model receives a relatively higher weight when calculating the average forecast.

More formally, a Bates–Granger [5] combined forecast F_h^{BG} of overnights/arrivals for non-residents/ residents for a forecast horizon h ($h = 3, 6$) is calculated as follows:

$$F_h^{BG} = \sum_{m=1}^{M} \frac{1/MSE_h^m}{\sum_{m=1}^{M}\left(1/MSE_h^m\right)} F_h^m \tag{8}$$

where F_h^m in Equation (8) is a forecast value produced by one of the single m ($m = 1, \ldots, M$) competing single forecasting models, and MSE_h^m denotes the corresponding mean square error.

5. Results

The forecasting models were assessed based on the comparison of their ex-post out-of-sample forecasting accuracy in terms of the root mean square error (RMSE), the mean absolute error (MAE), and the mean absolute percentage error (MAPE). The averaged (or combined) forecasts were then treated the same way as the forecast values produced by the individual forecasting models, which means that the same error measures are also calculated for them.

To evaluate the forecasting accuracy of the different models, we ranked the values yielded by the various forecasting accuracy measures employed and summated the scores: this procedure allows for the interpretation that the forecasting model with the lowest total score delivers the best forecasting accuracy (see as an example the results for the December 2017 report in Tables A1–A8 in Appendix A). In the next step, we compared the total scores added over eight report periods (March 2016, June 2016, September 2016, December 2016, March 2017, June 2017, September 2017, December 2017) and over three forecasting accuracy measures (RMSE, MAE, and MAPE) to determine whether or not the combined forecasting methods outperformed the single forecasting models (see Tables 1 and 2). These tables present an evaluation by forecast horizon separated between non-residents and residents, as well as an overall rank. In addition, one goal of the commissioned study was to give an overall recommendation across forecasting accuracy measures, report periods, and forecasting horizons. Therefore, presenting an overall rank became necessary as well.

Comparing the forecasting performance between March 2015 and August 2017, incorporating eight different forecasting situations as well as different macroeconomic environments, the combined forecasts with Bates–Granger weights continuously deliver the most accurate forecasts (see Tables 1 and 2). Generally speaking, this is the case for arrivals, overnights, all forecast horizons, and tourist types. However, in one case, the six months ahead forecasts of overnights by non-residents, the combined forecast approach based on Bates–Granger weights ranked second behind the ARIMA (2,1,0) model, thus failing to perform best as it had in all other cases, although the differences in the rank totals were so small that both methods could be considered as sharing the first rank.

Table 1. Summary ranking of the overall results for arrivals forecasts. Source: European Commission (2017) and own calculations. Boldface indicates the single forecasting model(s)/forecast combination technique(s) characterized by the lowest sum of ranks.

Arrivals	3 Months Non-Residents	6 Months Non-Residents	3 Months Residents	6 Months Residents	Sum	Overall Rank
ARIMA (2,1,0)	109	117	119	161	506	4
REGARIMA AR(1)	194	181	180	144	699	9
ETS (AIC)	189	199	140	125	653	8
ETS (BIC)	194	204	168	164	730	10
REGARIMA + Google Trends (EU_trends)	124	72	115	109	420	3
REGARIMA + Google Trends (EU_flights)	132	122	150	158	562	6
REGARIMA + Google Trends (EU_ hotels)	117	112	151	152	532	5
REGARIMA + Google Trends (EU_travel)	145	106	154	173	578	7
Combined forecasts (uniform weights)	59	118	91	76	344	2
Combined forecasts (Bates–Granger weights)	**57**	**62**	**52**	**54**	**225**	**1**

Table 2. Summary ranking of the overall results for overnights forecasts. Source: European Commission (2017) and own calculations. Boldface indicates the single forecasting model(s)/forecast combination technique(s) characterized by the lowest sum of ranks.

Overnights	3 Months Non-Residents	6 Months Non-Residents	3 Months Residents	6 Months Residents	Sum	Overall Rank
ARIMA (2,1,0)	87	55	138	78	358	3
REGARIMA AR(1)	132	130	122	103	487	4
ETS (AIC)	168	211	104	106	589	6
ETS (BIC)	164	196	117	106	583	5
REGARIMA + Google Trends (EU_trends)	148	127	168	165	608	7
REGARIMA + Google Trends (EU_flights)	173	150	129	154	606	9
REGARIMA + Google Trends (EU_ hotels)	171	122	157	168	618	8
REGARIMA + Google Trends (EU_travel)	165	130	165	164	624	10
Combined forecasts (uniform weights)	63	140	86	101	390	2
Combined forecasts (Bates–Granger weights)	**49**	**56**	**39**	**59**	**203**	**1**

In the competition for the top three ranks, the combined forecasts with Bates–Granger weights ranked first seven times and came second once. The ARIMA models ranked first once, tightly beating the weighted combined forecasts, ranked second once, and ranked third twice. The combined forecasts with uniform weights achieved the second rank five times and the third rank once. The REGARIMA models with Google Trends indices for the European Union scored one second rank and two thirds. The ETS (AIC) models and REGARIMA models with Google Trends indices for hotels and travel each achieved the third rank a single time.

When analyzing the overall ranks of the two employed forecasting combination techniques separately for the eight report periods, combined forecasts based on Bates–Granger weights achieved the lowest sum of ranks in 22 out of 64 possible cases, whereas combined forecasts based on uniform weights achieved the lowest sum of ranks in 16 cases (detailed results are available from the authors on request). Thus, for approximately 60% of all cases and when only considering best ranks, averaging over the results of the individual forecasting models was definitely worthwhile. It should also be noted that none of the individual forecasting models performed extremely poorly and that the results were quite similar across report periods. For the given sample, minimizing the impact of individual forecasting models performing slightly poorly in terms of assigning a lower weight when calculating the Bates–Granger combined forecast proved to be sufficient. Consequently, more complex approaches (e.g., using a formal screening procedure based on statistical criteria such as a forecast encompassing test) was not necessary (and would also have been at odds with the simplicity requirement for the methodology).

6. Conclusions

This study compared the accuracy of individual and combined approaches to forecasting tourism demand for the total European Union. The evaluation of the forecasting accuracies was performed recursively for eight periods spanning two years in order to check the stability of the outcomes during a changing macroeconomic environment. The analysis of the out-of-sample forecasts for arrivals and overnights showed that forecast combinations taking the historical forecasting performance of individual approaches such as Autoregressive Integrated Moving Average (ARIMA) models, REGARIMA models with different trend variables, and Error Trend Seasonal (ETS) models into account deliver the best results.

The analysis of the three months ahead and six months ahead out-of-sample forecasts for arrivals and overnights (non-residents, residents) showed that the VACO method was clearly more accurate, followed by the ARIMA approach, and the uniformly weighted combined forecasting method (SAC). The results and their stability over a two-year observation period demonstrate that taking the historical forecasting performance into account contributes to a significant improvement of forecasting accuracy and is recommended for practical application.

One particular advantage of the SAC and VACO methods in addition to their excellent performance is that they can be easily implemented by practitioners, which typically is not the case for more complex forecast combination techniques such as multiple encompassing tests. This simplicity also holds for the employed single forecasting models, which are typically part of any modern statistics and econometrics software. One further advantage is the ready availability of the different trend variables employed as exogenous variables in the REGARIMA models, such as HP trends and the various Google Trends indices.

Limitations of our study are that we used very simple models for practical reasons and did not test whether the introduction of complex models in the forecasting competition would have changed the rank orders in terms of the forecasting accuracies. In line with the general need to improve forecasting accuracy, discussions of complexity have also become increasingly relevant in the academic literature, although building complex approaches is at odds with scientific principles that advocate simplicity. On the other hand, a way to use complex techniques in order to improve forecasting accuracy is also the combination of the forecasts of individual forecasting models, as employed in this study, since combined methods minimize the risk of high inaccuracy by "averaging out" the weaknesses of the single forecasting models. They are also capable of introducing adjustments and additional information, which can balance out measurement errors and thereby affect forecasting power.

Future research efforts could concentrate on developing forecasting models for the single European Union countries and testing if the forecasting accuracy performances of the different methods stay stable across countries or if significant differences can be detected. For practical reasons, such country-level approaches could even be more useful for governments and national tourism boards. Another research endeavor could investigate the sensitivity of the forecasting performances of the different methods in terms of different data frequencies (e.g., quarters instead of months).

Author Contributions: Conceptualization, U.G., I.Ö., and E.S.; methodology, U.G., I.Ö., and E.S.; software, U.G., I.Ö., and E.S.; validation, U.G., I.Ö., and E.S.; formal analysis, U.G., I.Ö., and E.S.; investigation, U.G., I.Ö., and E.S.; resources, I.Ö.; data curation, I.Ö.; writing—original draft preparation, U.G., I.Ö., and E.S.; writing—review and editing, U.G.; visualization, U.G.; supervision, E.S.; project administration, E.S.; funding acquisition, E.S. All authors have read and agreed to the published version of the manuscript.

Funding: This research was funded by European Commission (Directorate General for Internal Market, Industry, Entrepreneurship & SMEs – Directorate F: Innovation & Manufacturing industries) under the title "Statistical Report on Tourism Accommodation Establishments", tender number GRO/SME/15/C/N125C.

Acknowledgments: The authors would like to thank the two anonymous reviewers of this journal, whose comments and suggestions could greatly improve the quality of the manuscript.

Conflicts of Interest: The authors declare no conflict of interest.

Appendix A

Appendix A.1. Arrivals (December 2017 Results)

Table A1. Forecasting accuracy (December 2017): arrivals, non-residents, three months ahead. Source: European Commission (2017) and own calculations. Boldface indicates the single forecasting model(s)/forecast combination technique(s) characterized by the lowest sum of ranks.

3 Months Forecasting Horizon: Non-Residents	RMSE	Rank	MAE	Rank	MAPE	Rank	Sum of Ranks
ARIMA (2,1,1)	1,962,919.00	5	1,389,331.00	5	3.78	5	15
REGARIMA MA(1)	1,968,167.00	6	1,558,767.00	8	4.63	9	23
M, MD, A (AIC)	674,491.76	2	669,504.32	2	1.97	2	6
M, N, A (BIC)	1,947,636.44	4	1,773,359.65	10	4.98	10	24
REGARIMA (2,1,1) + Google Trends (EU)	2,142,712.00	10	1,618,392.00	9	4.42	8	27
REGARIMA (2,1,1) + Google Trends (EU_flights)	2,031,774.00	8	1,451,146.00	6	3.95	6	20
REGARIMA (2,1,1) + Google Trends (EU_hotels)	1,981,988.00	7	1,368,884.00	4	3.69	4	15
REGARIMA (2,1,1) + Google Trends (EU_travel)	2,058,870.00	9	1,493,207.00	7	4.07	7	23
Combined forecasts (uniform weights)	1,103,099.45	3	977,291.59	3	2.73	3	9
Combined forecasts (Bates–Granger weights)	**624,338.87**	**1**	**625,290.19**	**1**	**1.81**	**1**	**3**

Table A2. Forecasting accuracy (December 2017): arrivals, non-residents, six months ahead. Source: European Commission (2017) and own calculations. Boldface indicates the single forecasting model(s)/forecast combination technique(s) characterized by the lowest sum of ranks.

6 Months Forecasting Horizon: Non-Residents	RMSE	Rank	MAE	Rank	MAPE	Rank	Sum of Ranks
ARIMA (2,1,1)	1,491,331.00	7	1,339,141.00	7	4.95	7	21
REGARIMA MA(1)	2,447,610.00	8	1,874,815.00	8	7.01	8	24
M, MD, A (AIC)	3,869,291.14	9	3,180,772.49	10	10.07	10	29
M, N, A (BIC)	3,274,107.22	10	2,567,872.69	9	7.90	9	28
REGARIMA (2,1,1) + Google Trends (EU)	1,465,033.00	5	1,231,018.00	3	4.50	4	12
REGARIMA (2,1,1) + Google Trends (EU_flights)	1,460,824.00	4	1,280,448.00	6	4.69	6	16
REGARIMA (2,1,1) + Google Trends (EU_hotels)	1,397,517.00	2	1,234,282.00	4	4.47	3	9
REGARIMA (2,1,1) + Google Trends (EU_travel)	1,467,741.00	6	1,274,591.00	5	4.67	5	16
Combined forecasts (uniform weights)	**1,412,399.45**	**3**	**996,892.74**	**1**	**3.78**	**1**	**5**
Combined forecasts (Bates–Granger weights)	**1,344,569.48**	**1**	**1,199,343.49**	**2**	**4.45**	**2**	**5**

Table A3. Forecasting accuracy (December 2017): arrivals, residents, three months ahead. Source: European Commission (2017) and own calculations. Boldface indicates the single forecasting model(s)/forecast combination technique(s) characterized by the lowest sum of ranks.

3 Months Forecasting Horizon: Residents	RMSE	Rank	MAE	Rank	MAPE	Rank	Sum of Ranks
ARIMA (2,1,0)	1,460,675.00	6	1,114,304.00	7	2.47	7	20
REGARIMA AR(1)	1,769,949.00	10	1,645,885.00	10	3.86	10	30
M, M, M (AIC)	784,126.59	2	717,596.63	2	1.71	2	6
M, N, M (BIC)	**625,913.79**	**1**	**505,672.84**	**1**	**1.22**	**1**	**3**
REGARIMA (2,1,0) + Google Trends (EU)	1,567,547.00	9	1,037,070.00	5	2.27	5	19
REGARIMA (2,1,0) + Google Trends (EU_flights)	1,464,579.00	8	1,127,056.00	9	2.50	9	26
REGARIMA (2,1,0) + Google Trends (EU_hotels)	1,456,503.00	5	1,120,557.00	8	2.48	8	21
REGARIMA (2,1,0) + Google Trends (EU_travel)	1,462,131.00	7	1,076,862.00	6	2.38	6	19
Combined forecasts (uniform weights)	1,029,620.56	3	796,857.07	3	1.77	3	9
Combined forecasts (Bates–Granger weights)	1,211,760.77	4	920,354.25	4	2.04	4	12

Table A4. Forecasting accuracy (December 2017): arrivals, residents, six months ahead. Source: European Commission (2017) and own calculations. Boldface indicates the single forecasting model(s)/forecast combination technique(s) characterized by the lowest sum of ranks.

6 Months Forecasting Horizon: Residents	RMSE	Rank	MAE	Rank	MAPE	Rank	Sum of Ranks
ARIMA (2,1,0)	1,803,223.00	8	943,557.00	9	2.41	9	26
REGARIMA AR(1)	1,665,416.00	10	1,482,656.00	10	3.84	10	30
M, A, M (AIC)	556,627.45	2	476,424.56	2	1.20	2	6
M, N, M (BIC)	**501,694.40**	**1**	**394,697.50**	**1**	**1.01**	**1**	**3**
REGARIMA (2,1,0) + Google Trends (EU)	1,073,848.00	6	849,060.20	5	2.16	5	16
REGARIMA (2,1,0) + Google Trends (EU_flights)	1,075,211.00	7	940,386.60	7	2.41	8	22
REGARIMA (2,1,0) + Google Trends (EU_hotels)	982,754.80	5	861,745.50	6	2.18	6	17
REGARIMA (2,1,0) + Google Trends (EU_travel)	1,085,191.00	9	921,788.10	8	2.36	7	24
Combined forecasts (uniform weights)	818,608.72	4	726,462.04	4	1.87	4	12
Combined forecasts (Bates–Granger weights)	583,003.73	3	531,839.88	3	1.36	3	9

Appendix .1. Overnights (December 2017 Results)

Table A5. Forecasting accuracy (December 2017): overnights, non-residents, three months ahead. Source: European Commission (2017) and own calculations. Boldface indicates the single forecasting model(s)/forecast combination technique(s) characterized by the lowest sum of ranks.

3 Months Forecasting Horizon: Non-Residents	RMSE	Rank	MAE	Rank	MAPE	Rank	Sum of Ranks
ARIMA (2,1,0)	6,605,540.00	4	4,387,028.00	3	5.00	4	11
REGARIMA AR(1)	6,320,525.00	3	4,001,776.00	2	3.21	2	7
M, MD, A (AIC)	9,737,534.88	6	12,199,817.86	6	6.75	5	17
M, N, A (BIC)	8,645,659.56	5	11,330,813.11	5	8.86	6	16
REGARIMA (2,2,0) + Google Trends (EU)	16,916,357.00	10	16,700,366.00	10	13.32	10	30
REGARIMA (2,2,0) + Google Trends (EU_flights)	15,693,427.00	8	15,485,019.00	8	12.36	8	24
REGARIMA (2,2,0) + Google Trends (EU_hotels)	15,068,359.00	7	14,830,699.00	7	11.84	7	21
REGARIMA (2,2,0) + Google Trends (EU_travel)	15,903,711.00	9	15,693,577.00	9	12.53	9	27
Combined forecasts (uniform weights)	6,176,397.10	2	5,631,492.19	4	4.53	3	9
Combined forecasts (Bates–Granger weights)	**2,166,930.13**	**1**	**1,883,165.39**	**1**	**1.44**	**1**	**3**

Table A6. Forecasting accuracy (December 2017): overnights, non-residents, six months ahead. Source: European Commission (2017) and own calculations. Boldface indicates the single forecasting model(s)/forecast combination technique(s) characterized by the lowest sum of ranks.

6 Months Forecasting Horizon: Non-Residents	RMSE	Rank	MAE	Rank	MAPE	Rank	Sum of Ranks
ARIMA (2,1,0)	**5,465,725.00**	**1**	**4,463,842.00**	**1**	**5.09**	**1**	**3**
REGARIMA AR(1)	7,197,605.00	2	5,660,215.00	2	6.97	3	7
M, MD, A (AIC)	15,570,760.51	9	13,704,608.24	9	12.80	9	27
M, N, A (BIC)	15,918,015.56	10	14,012,547.74	10	13.07	10	30
REGARIMA (2,2,0) + Google Trends (EU)	12,370,566.00	5	10,752,681.00	5	10.27	5	15
REGARIMA (2,2,0) + Google Trends (EU_flights)	13,029,867.00	7	11,283,221.00	7	10.69	7	21
REGARIMA (2,2,0) + Google Trends (EU_hotels)	14,427,259.00	8	12,459,483.00	8	11.64	8	24
REGARIMA (2,2,0) + Google Trends (EU_travel)	12,820,154.00	6	11,109,223.00	6	10.55	6	18
Combined forecasts (uniform weights)	11,420,444.83	4	9,918,838.29	4	9.50	4	12
Combined forecasts (Bates–Granger weights)	7,814,152.04	3	6,596,603.17	3	6.95	2	8

Table A7. Forecasting accuracy (December 2017): overnights, residents, three months ahead. Source: European Commission (2017) and own calculations. * AIC and BIC selected the same model. Boldface indicates the single forecasting model(s)/forecast combination technique(s) characterized by the lowest sum of ranks.

3 Months Forecasting Horizon: Residents	RMSE	Rank	MAE	Rank	MAPE	Rank	Sum of Ranks
ARIMA (2,1,0)	4,646,991.00	5	4,386,662.00	5	4.02	5	15
REGARIMA AR(1)	3,202,677.00	3	2,468,029.00	3	2.55	3	9
A, N, M (AIC)*	2,255,395.99	2	1,859,448.54	2	1.90	2	6
A, N, M (BIC)*	2,255,395.99	2	1,859,448.54	2	1.90	2	6
REGARIMA (2,2,0) + Google Trends (EU)	8,644,081.00	9	7,432,814.00	9	6.59	9	27
REGARIMA (2,2,0) + Google Trends (EU_flights)	6,884,342.00	7	5,586,683.00	7	4.91	7	21
REGARIMA (2,2,0) + Google Trends (EU_hotels)	6,347,254.00	6	5,179,781.00	6	4.55	6	18
REGARIMA (2,2,0) + Google Trends (EU_travel)	7,349,120.00	8	6,081,357.00	8	5.36	8	24
Combined forecasts (uniform weights)	4,024,454.10	4	3,681,202.32	4	3.32	4	12
Combined forecasts (Bates–Granger weights)	**1,880,723.47**	**1**	**1,486,759.71**	**1**	**1.51**	**1**	**3**

Table A8. Forecasting accuracy (December 2017): overnights, residents, six months ahead. Source: European Commission (2017) and own calculations. * AIC and BIC selected the same model. Boldface indicates the single forecasting model(s)/forecast combination technique(s) characterized by the lowest sum of ranks.

6 Months Forecasting Horizon: Residents	RMSE	Rank	MAE	Rank	MAPE	Rank	Sum of Ranks
ARIMA (2,1,0)	3,157,904.00	3	2,598,295.00	3	3.17	3	9
REGARIMA AR(1)	5,375,422.00	4	4,629,242.00	4	6.04	5	13
A, N, M (AIC)*	**2,271,676.09**	**1**	**1,851,199.77**	**1**	**2.22**	**1**	**3**
A, N, M (BIC)*	**2,271,676.09**	**1**	**1,851,199.77**	**1**	**2.22**	**1**	**3**
REGARIMA (2,2,0) + Google Trends (EU)	9,454,403.00	6	8,432,516.00	6	8.98	6	18
REGARIMA (2,2,0) + Google Trends (EU_flights)	10,433,326.00	8	9,174,597.00	8	9.63	8	24
REGARIMA (2,2,0) + Google Trends (EU_hotels)	11,910,297.00	9	10,428,86.00	9	10.88	9	27
REGARIMA (2,2,0) + Google Trends (EU_travel)	10,027,850.00	7	8,855,383.00	7	9.34	7	21
Combined forecasts (uniform weights)	5,684,099.00	5	5,174,079.48	5	5.71	4	14
Combined forecasts (Bates–Granger weights)	2,319,373.77	2	1,971,291.24	2	2.43	2	6

References

1. Li, G.; Song, H.; Witt, S.F. Recent Developments in Econometric Modeling and Forecasting. *J. Travel Res.* **2005**, *44*, 82–99. [CrossRef]
2. Green, K.C.; Armstrong, J.S. Simple versus Complex Forecasting: The Evidence. *J. Bus. Res.* **2015**, *68*, 1678–1685. [CrossRef]
3. Shen, S.; Li, G.; Song, H. An Assessment of Combining Tourism Demand Forecasts over Different Time Horizons. *J. Travel Res.* **2008**, *47*, 197–207. [CrossRef]
4. Makridakis, S.; Wheelwright, S.C.; Hyndman, R.J. *Forecasting: Methods and Applications*, 3rd ed.; John Wiley&Sons Inc.: Hoboken, NJ, USA, 1998.
5. Bates, J.M.; Granger, C.W.J. The Combination of Forecasts. *J. Oper. Res. Soc.* **1969**, *20*, 451–468. [CrossRef]
6. Clemen, R.T. Combining forecasts: A review and annotated bibliography. *Int. J. Forecast.* **1989**, *5*, 559–583. [CrossRef]
7. Andrawis, R.; Atiya, A.F.; El-Shishiny, H.E.E.-D. Combination of long term and short term forecasts, with application to tourism demand forecasting. *Int. J. Forecast.* **2011**, *27*, 870–886. [CrossRef]
8. Calantone, R.J.; Di Benedetto, A.; Bojanic, D.C. Multimethod forecasts for tourism analysis. *Ann. Tour. Res.* **1988**, *15*, 387–406. [CrossRef]
9. Chu, F.-L. Forecasting tourism: A combined approach. *Tour. Manag.* **1998**, *19*, 515–520. [CrossRef]
10. Fritz, R.G.; Brandon, C.; Xander, J. Combining Time Series and Econometric Forecasts of Tourism Activity. *Ann. Tour. Res.* **1984**, *11*, 219–229. [CrossRef]
11. Gasmi, A. Combination Forecasts of International Demand for Tourism in Tunisia. *J. Quant. Econ.* **2014**, *12*, 94–109.
12. Gunter, U.; Önder, I. Forecasting city arrivals with Google Analytics. *Ann. Tour. Res.* **2016**, *61*, 199–2122. [CrossRef]
13. Oh, C.-O.; Morzuch, B.J. Evaluating Time-Series Models to Forecast the Demand for Tourism in Singapore. *J. Travel Res.* **2005**, *43*, 404–413. [CrossRef]
14. European Commission. *Statistical Report on Tourism Accommodation Establishments*; European Commission (EC): Brussels, Belgium, 2017.
15. Eurostat. Statistics Explained—Tourism Statistics. 2020. Available online: https://ec.europa.eu/eurostat/statistics-explained/index.php?title=Tourism_statistics (accessed on 22 June 2020).
16. Hyndman, R.J. A Brief History of Forecasting Competitions. 2019. Available online: https://robjhyndman.com/papers/forecasting-competitions.pdf. (accessed on 22 June 2020).
17. Reid, D. A Comparative Study of Time Series Prediction Techniques on Economic Data. Ph.D. Thesis, University of Nottingham, Nottingham, UK, 1969.
18. Newbold, P.; Granger, C.W.J. Experience with Forecasting Univariate Time Series and the Combination of Forecasts. *J. R. Stat. Soc. Ser. A (Gen.)* **1974**, *137*, 131. [CrossRef]
19. Makridakis, S.; Andersen, A.; Carbone, R.; Fildes, R.; Hibon, M.; Lewandowski, R.; Newton, J.; Parzen, E.; Winkler, R. The accuracy of extrapolation (time series) methods: Results of a forecasting competition. *J. Forecast.* **1982**, *1*, 111–153. [CrossRef]
20. Makridakis, S.; Chatfield, C.; Hibon, M.; Lawrence, M.; Mills, T.; Ord, K.; Simmons, L.F. The M2-competition: A real-time judgmentally based forecasting study. *Int. J. Forecast.* **1993**, *9*, 5–22. [CrossRef]
21. Makridakis, S.; Hibon, M. The M3-Competition: Results, conclusions and implications. *Int. J. Forecast.* **2000**, *16*, 451–476. [CrossRef]
22. Makridakis, S.; Spiliotis, E.; Assimakopoulos, V. The M4 Competition: 100,000 time series and 61 forecasting methods. *Int. J. Forecast.* **2020**, *36*, 54–74. [CrossRef]
23. Gershenfeld, N.A.; Weigend, A.S. (Eds.) The future of time series. In *Time Series Prediction: Forecasting the Future and Understanding the Past*; Westview Press: Boulder, CO, USA, 1993; pp. 1–70.
24. Crone, S.F.; Hibon, M.; Nikolopoulos, K. Advances in forecasting with neural networks? Empirical evidence from the NN3 competition on time series prediction. *Int. J. Forecast.* **2011**, *27*, 635–660. [CrossRef]
25. Hong, T.; Pinson, P.; Fan, F. Global Energy Forecasting Competition 2012. *Int. J. Forecast.* **2014**, *30*, 357–363. [CrossRef]
26. Hong, T.; Pinson, P.; Fan, S.; Zareipour, H.; Troccoli, A.; Hyndman, R.J. Probabilistic energy forecasting: Global Energy Forecasting Competition 2014 and beyond. *Int. J. Forecast.* **2016**, *32*, 896–913. [CrossRef]

27. Song, H.; Witt, S.F.; Wong, K.F.; Wu, D.C. An Empirical Study of Forecast Combination in Tourism. *J. Hosp. Tour. Res.* **2009**, *33*, 3–29. [CrossRef]
28. Shen, S.; Li, G.; Song, H. Combination forecasts of International tourism demand. *Ann. Tour. Res.* **2011**, *38*, 72–89. [CrossRef]
29. Granger, C.W.J.; Ramanathan, R. Improved methods of combining forecasts. *J. Forecast.* **1984**, *3*, 197–204. [CrossRef]
30. Cang, S. A non-linear tourism demand forecast combination model. *Tour. Econ.* **2011**, *17*, 5–20. [CrossRef]
31. Song, H.; Wong, K.K.; Witt, S.F. Assessing the impact of forecast combination on tourism demand forecasting accuracy. In *The Routledge Handbook of Tourism Research*; Routledge: Abingdon, UK, 2012; pp. 93–109.
32. Costantini, M.; Gunter, U.; Kunst, R.M. Forecast Combinations in a DSGE-VAR Lab. *J. Forecast.* **2016**, *36*, 305–324. [CrossRef]
33. Athanasopoulos, G.; Hyndman, R.J.; Song, H.; Wu, D.C. The tourism forecasting competition. *Int. J. Forecast.* **2011**, *27*, 822–844. [CrossRef]
34. Clemen, R.T. Combining Overlapping Information. *Manag. Sci.* **1987**, *33*, 373–380. [CrossRef]
35. Chong, Y.Y.; Hendry, D.F. Econometric Evaluation of Linear Macro-Economic Models. *Rev. Econ. Stud.* **1986**, *53*, 671. [CrossRef]
36. Diebold, F.X. Forecast combination and encompassing: Reconciling two divergent literatures. *Int. J. Forecast.* **1989**, *5*, 589–592. [CrossRef]
37. Pesaran, M.H.; Timmermann, A. Selection of estimation window in the presence of breaks. *J. Econ.* **2007**, *137*, 134–161. [CrossRef]
38. Stock, J.H.; Watson, M.W. Combination forecasts of output growth in a seven-country data set. *J. Forecast.* **2004**, *23*, 405–430. [CrossRef]
39. Hendry, D.F.; Clements, M.P. Pooling of forecasts. *Econ. J.* **2004**, *7*, 1–31. [CrossRef]
40. Timmerman, A. Forecast combination. In *Handbook of Economic Forecasting*; Elliott, G., Granger, C.W.J., Timmerman, A., Eds.; Elsevier: Amsterdam, The Netherlands, 2006.
41. Palm, F.C.; Zellner, A. To combine or not to combine? issues of combining forecasts. *J. Forecast.* **1992**, *11*, 687–701. [CrossRef]
42. Elliott, G.; Timmermann, A. Optimal forecast combinations under general loss functions and forecast error distributions. *J. Econ.* **2004**, *122*, 47–79. [CrossRef]
43. Enders, W. *Applied Econometric Time Series*; John Wiley & Sons Inc.: Hoboken, NJ, USA, 2014.
44. EViews. *EViews 8 User's Guide I and II*; IHS Global Inc.: Irvine, CA, USA, 2014.
45. Hyndman, R.J.; Koehler, A.B.; Ord, J.K.; Snyder, R.D. *Forecasting with Exponential Smoothing: The State Space Approach*; Springer: Berlin/Heidelberg, Germany, 2008.
46. Hodrick, R.J.; Prescott, E.C. Postwar U.S. Business Cycles: An Empirical Investigation. *J. Money Credit Bank.* **1997**, *29*, 1–16. [CrossRef]
47. Huang, X.; Zhang, L.; Ding, Y. The Baidu Index: Uses in predicting tourism flows—A case study of the Forbidden City. *Tour. Manag.* **2017**, *58*, 301–306. [CrossRef]
48. Wasserstein, R.L.; Lazar, N.A. The ASA's statement on p-values: Context, process, and purpose. *Am. Stat.* **2016**, *70*, 129–133. [CrossRef]
49. Gunter, U.; Önder, I.; Smeral, E. Scientific value of econometric tourism demand studies. *Ann. Tour. Res.* **2019**, *78*, 102738. [CrossRef]
50. Wasserstein, R.L.; Schirm, A.L.; Lazar, N.A. Moving to a world beyond 'p <0.05'. *Am. Stat.* **2019**, *73*, 1–19.
51. Hyndman, R.J.; Koehler, A.B.; Snyder, R.D.; Grose, S. A state space framework for automatic forecasting using exponential smoothing methods. *Int. J. Forecast.* **2002**, *18*, 439–454. [CrossRef]
52. Hyndman, R.J.; Athanasopoulos, G. *Forecasting: Principles and Practice*, 2nd ed.; OTexts: Melbourne, Australia, 2018. Available online: https://otexts.com/fpp2/ (accessed on 14 November 2019).

Article

Benchmarking Real-Time Streamflow Forecast Skill in the Himalayan Region

Ganesh R. Ghimire [1,*], Sanjib Sharma [2], Jeeban Panthi [3], Rocky Talchabhadel [4], Binod Parajuli [5], Piyush Dahal [6] and Rupesh Baniya [7]

1 IIHR-Hydroscience and Engineering, The University of Iowa, Iowa City, IA 52242, USA
2 Earth and Environmental Systems Institute, The Pennsylvania State University, University Park, PA 16801, USA; sanjibsharma66@gmail.com
3 Department of Geosciences, University of Rhode Island, Kingston, RI 02881, USA; Panthijeeban@gmail.com
4 Disaster Prevention Research Institute, Kyoto University, Fushimi-ku, Kyoto 612-8235, Japan; rocky.ioe@gmail.com
5 Department of Hydrology and Meteorology, Ministry of Energy, Water Resources and Irrigation, Kathmandu 44600, Nepal; binodparaj@gmail.com
6 The Small Earth Nepal, Kathmandu 44600, Nepal; piyush.dahal@gmail.com
7 Institute of Engineering, Pulchowk Campus, Tribhuvan University, Lalitpur 44700, Nepal; rupesh.baniya480@gmail.com
* Correspondence: ganeshghimire1986@gmail.com; Tel.: +1-618-340-2613

Received: 15 June 2020; Accepted: 6 July 2020; Published: 8 July 2020

Abstract: Improving decision-making in various areas of water policy and management (e.g., flood and drought preparedness, reservoir operation and hydropower generation) requires skillful streamflow forecasts. Despite the recent advances in hydrometeorological prediction, real-time streamflow forecasting over the Himalayas remains a critical issue and challenge, especially with complex basin physiography, shifting weather patterns and sparse and biased in-situ hydrometeorological monitoring data. In this study, we demonstrate the utility of low-complexity data-driven persistence-based approaches for skillful streamflow forecasting in the Himalayan country Nepal. The selected approaches are: (1) simple persistence, (2) streamflow climatology and (3) anomaly persistence. We generated the streamflow forecasts for 65 stream gauge stations across Nepal for short-to-medium range forecast lead times (1 to 12 days). The selected gauge stations were monitored by the Department of Hydrology and Meteorology (DHM) Nepal, and they represent a wide range of basin size, from ~17 to ~54,100 km^2. We find that the performance of persistence-based forecasting approaches depends highly upon the lead time, flow threshold, basin size and flow regime. Overall, the persistence-based forecast results demonstrate higher forecast skill in snow-fed rivers over intermittent ones, moderate flows over extreme ones and larger basins over smaller ones. The streamflow forecast skill obtained in this study can serve as a benchmark (reference) for the evaluation of many operational forecasting systems over the Himalayas.

Keywords: Himalayan region; streamflow forecast verification; persistence; snow-fed rivers; intermittent rivers

---

1. Introduction

Skillful streamflow forecasts are critically important in improving decision-making in water-related policy and management (e.g., flood and drought preparedness, reservoir operation and hydropower generation). Real-time streamflow forecasting over the Himalayas remains a critical challenge. (e.g., [1,2]) because of complex basin physiography, shifting weather patterns and sparse distribution of hydrometeorological monitoring stations. The verification of streamflow forecasts can provide insight into hydrologic forecasting by elucidating the level of forecast skill that can be reasonably expected from the forecasting system [3]. In this study, we demonstrate the utility of a persistence-based approach to benchmark forecast skill of the operational streamflow forecasting system in the Himalayan country Nepal.

Nepal is an important part of the Hindu Kush Himalayas (HKH), which are referred to as the water towers of Asia due to the largest concentration of snow and glaciers outside the two poles [4–7]. The snow cover and glaciers make significant contributions to the hydrology of glacierized basins. Khadka et al. [5] showed, in the simulation study of Tamakoshi basin in eastern Nepal for the years 2000–2009, that snowmelt contributes about 18% of the annual runoff. Further, they showed that the snowmelt represents about 17% of the runoff in the summer season (June to September) while about 25% in the spring season, when the streamflow in these rivers are low. The study by Bhattarai and Regmi [8] in the Langtang river basin in central Nepal shows similar numbers on the contribution of snowmelt to the runoff (~19% on the annual runoff). Dhami et al. [9] used the Soil and Water Assessment Tool and snow-melt runoff model in the Karnali river basin in western Nepal to simulate components of water balance. They reported that about 12% of annual runoff is contributed by the snowmelt, while about 29% by the groundwater base flow.

In Nepal, the Department of Hydrology and Meteorology (DHM) is responsible for monitoring the streamflow at rivers, providing real-time river watch, and developing decision support systems. At present, the real-time river watch system at DHM generates siren alerts at critical stations leveraging the internet and mobile devices for alert delivery [10]. In addition, the community-based early warning system (CBEWS) has been integrated by the early warning system managed by the DHM with the aid of intergovernmental organizations, UN organizations and international non-governmental organizations [11]. CBEWS is inherently people-centric, helping communities use local resources and capacities for flood preparedness efforts [10–12]. The simplified data-based mechanistic forecast model was adopted by the DHM to cover the major river basins of Nepal that stem from the mountains to the Terai plains. The integration of the data-based mechanistic model with CBEWS provides an additional lead time of 3–5 h for larger basins and 1–3 h for smaller basins in addition to previous 2–3 h [13]. Owing to the major contribution of snow to the streamflow, particularly in the glacierized basins, persistence-based forecasts could serve as reference to evaluate the skill of the current operational forecasting system, which this study explores. Moreover, this approach could serve as a tool to guide the decision-making pertaining to flood preparedness in the short-medium forecast range.

Hydrologic forecasts are inherently uncertain (e.g., [3,14]). The uncertainty can originate from different sources, including shortcomings in forcing data (e.g., quantitative precipitation forecasts), model structure and parameters, as well as model initial conditions (e.g., [1,14–16]). In addition, complex hydrologic processes, such as snowmelt, sub-surface and flow routing, make it more difficult to produce skillful streamflow forecasts in the HKH region (e.g., [1,17]). Most countries in the HKH region are prone to floods (e.g., [1,18]); therefore, achieving skillful forecasts up to the medium range is of utmost importance for flood preparedness efforts. Because of such uncertainties and challenges, achieving skillful streamflow forecasts in this region requires skillful reference forecasting systems. For example, if a reference forecasting system with a low forecast skill is chosen, the operational streamflow (flood) forecasting system might depict higher forecast skill. It will essentially give a false sense of forecast skill associated with the operational system.

The persistence-based approach uses a notion of "tomorrow will be like today" (e.g., [19]). It is tied to the concept of "memory" of any analyzed system [20]: in our case, the basins. One would expect the notion of streamflow persistence to be more pertinent in the glacierized basins due to the relatively slow snowmelt process over time. Though persistence-based forecasts are in use as the reference in the hydrologic forecasting community, their utility across different hydroclimatic conditions and scales have not been fully explored.

Forecasters use various reference forecasts, such as simple persistence, streamflow climatology, a hydrological model fed with zero rainfall and a hydrological model fed with an ensemble of resampled historical rainfall [21], for operational forecast skill verification. Despite some inherent predictability of hydrologic systems due to coupling with the weather and climate system, streamflow shows some likelihood of repetition at seasonal to daily time scales, and hence some inherent predictability [19,20,22]. Den Dool [23] and Fraedrich and Ziehmann-Schlumbohm [24] iterate that only forecasts depicting skill better than persistence can handle the forecast of the time derivative. The degree of persistence, however, might demonstrate variation with location, time of the year and the system studied [23,25]. Den Dool [23] indicated that processes with a long-time scale show higher skill with persistence forecasts, which makes persistence a hard-to-beat method for many complex mechanistic models. Though some authors (e.g., [26]) attribute the streamflow persistence to the carryover storage of water in lakes, and below the land surface, the comprehensive evaluation of persistence-based forecasts was lacking until recently. Palash et al. [1], motivated by the concept of requisite simplicity, showed the utility of a simple linear flood-forecasting system for the Ganges, Brahmaputra and Meghna Rivers, using streamflow persistence as the mechanism of prediction. Recently, Ghimire and Krajewski [19] and Krajewski et al. [27] showed, through a comprehensive study of 140 mid-western agricultural watersheds in the United States, that persistence-based forecast skills show strong dependence with the basin scale, while weaker but non-negligible dependence with the properties of the river network.

In the context of a lack of comprehensive evaluation of persistence-based reference forecasts in the data-scarce Himalayan region, their quantitative assessment can provide the benchmark across scales for any operational forecasting system (e.g., DHM's forecasting system). This paper specifically explores the following questions: (1) what is the utility of data-driven persistence-based approaches for skillful streamflow forecasting in the Himalayan region? (2) Which forecast conditions, such as lead time, flow threshold, basin size and flow regime (e.g., perennial/snow-fed and intermittent), benefit potential increase in forecast skill? We organize the paper as follows: In Section 2, we discuss the materials and methods used in this study. Section 3 presents results of this study and Section 4 discusses these findings. In Section 5, we summarize and present some conclusions and limitations of this work.

2. Materials and Methods

2.1. Study Area and Data

Our domain of interest for this study was Nepal, which is in the central part of Himalayan mountain range. The total area of Nepal is about 147,520 km^2, located approximately between 80°03′–88°12′ E and 26°21′–30°26′ N (Figure 1). The topographic elevation ranges from 8848m (elevation of Mt. Everest) in the north to 70m in the south above mean sea level featuring diverse climatic conditions varying from polar to tropical [28]. Forest (39.1%), agriculture (29.8%), barren (10.7%) and snow/glaciers (8.2%) are the major land cover types in Nepal [29]. Most of the climate variability in Nepal is attributed to the high reliefs of river catchments e.g., [4]. The climate in Nepal is dominated by a southwestern monsoon (June–September) that originates from the Bay of Bengal and about 80% of precipitation in Nepal occurs only in the monsoon season [30]. The months of October–November occasionally experience post-monsoon rainfall, November–March typically remains dry, and April–May experiences pre-monsoon rainfall [4,5,9,31]. The summer monsoon is more active in the eastern part of Nepal as the monsoon enters and departs from that region, and the winter monsoon is more active in the western region because of the influence of western disturbances [28]. The average annual precipitation (1990–2010) ranges from ~270 mm to ~5500 mm [32]. The consequent runoff during the monsoon accounts for 70%–90% of the annual water balance, as shown in [33].

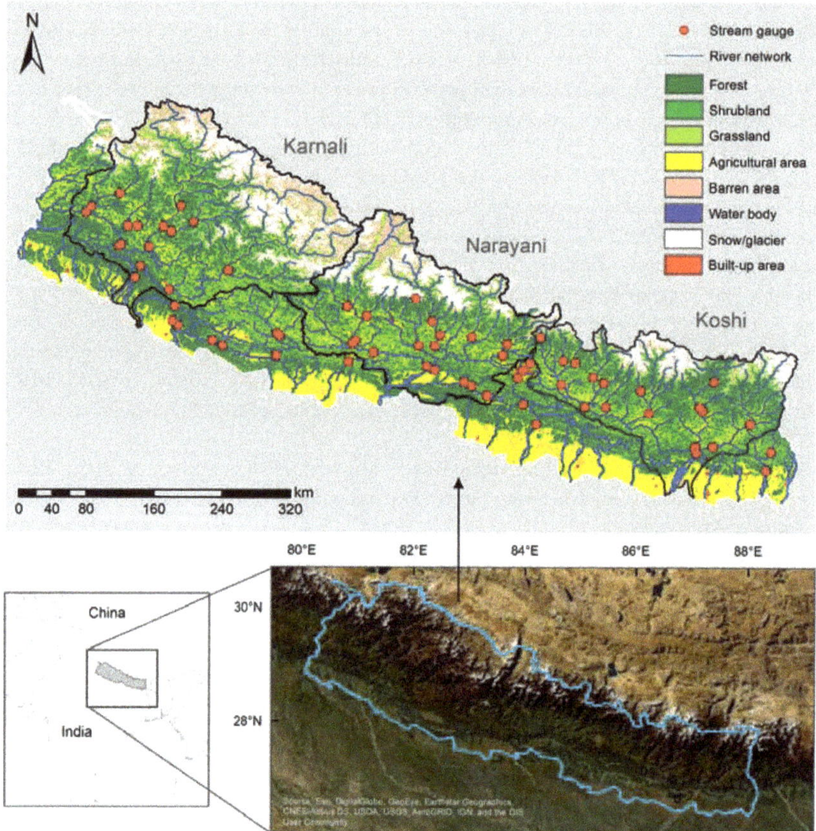

Figure 1. Location of the study domain, Nepal. The white patches in the terrain map (not to scale) depict the snow cover (Nepal Himalaya). The land cover is from Uddin et al. [29]. The red dots represent stream gauge stations monitored by the Department of Hydrology and Meteorology (DHM), Nepal.

Most rivers in Nepal drain from north to south and eventually to the Ganges River in India. The rivers in Nepal can be broadly categorized into five major river systems: Karnali, Narayani, Koshi, Mahakali, and the southern and Mahabharat rivers [33]. The former four river systems are combinations of rain- and snow-fed rivers, while the latter river system mostly comprise mid-size rivers, such as the Mechi, Kankai, Kamala, Bagmati, Babai and West Rapti, mostly fed by rainfall and characterized by frequent flash floods [33]. The Karnali, Gandaki and Koshi river systems highlighted in Figure 1 represent about two-thirds of the area of Nepal. These river systems show variability in their hydroclimatic conditions, in addition to the topographic characteristics such as average basin slope and snow cover. In this study, we used historic daily streamflow data at stream gauges stations (red dots in Figure 1) monitored by the DHM across Nepal. Daily streamflow data is the highest-resolution data publicly available from the DHM. It corresponds to the mean values of three readings recorded daily at 8 AM, 12 noon and 4 PM, local time. The streamflow data across stations span between 1962 and 2010, with varying record lengths. Thirty-eight stations have more than 30 years of historical daily streamflow records (Figure A1, Appendix A). We screened the stream gauge stations based on the availability criteria of at least 10 years of streamflow records and at least fair quality data categorized by DHM. The stations that are not shown within the three major river systems correspond to the rivers of the southern and Mahabharat river systems. In total, we used streamflow data from 65 unregulated stream gauge stations. The corresponding basin size varies from ~17 to ~54,100 km^2, with more than half of the stations monitoring catchments greater than 1000 km^2 (Figure A1, Appendix A). Note that the hydropower projects built in some of these rivers are run-off-river schemes, which typically do not store the water.

2.2. Experimental Design

Several persistence-based forecast approaches have been discussed in den Dool [19], Wu and Dickinson [23] and Ghimire and Krajewski [25]. Here, we present three of these approaches to generate the reference (benchmark) forecasts in the time domain and evaluate them; these are simple persistence, streamflow climatology and anomaly persistence. Our experimental setup mostly follows the methods outlined in Ghimire and Krajewski [19]. If t represents the time of observation of streamflow $Q(t)$, the forecast for lead time Δt is given by

$$Q(t + \Delta t) = Q(t) \tag{1}$$

Note that the forecasts are generated for the entire streamflow time-series in the records for 65 stream gauge stations in Nepal. While considering the entire time-series, we account for the interannual variability of streamflow.

Here, we use the "climatology" as the average of streamflow on record at time t. The corresponding forecast for lead time Δt is given by

$$Q(t + \Delta t) = \frac{\sum_{i=1}^{n} Q(t + \Delta t)}{n} \tag{2}$$

where $Q(t + \Delta t)$ refers to the streamflow at time $t + \Delta t$ from previous years of record and n refers to the number of years in record.

The anomaly persistence forecast scheme assumes that streamflow anomalies persist over the lead time Δt. The anomalies are computed with respect to the climatology. The anomalies at t and $t + \Delta t$ i.e., $Q'(t)$ and $Q'(t + \Delta t)$, respectively, are computed as

$$Q'(t) = Q(t) - clim(t) \tag{3}$$

$$Q'(t + \Delta t) = Q(t + \Delta t) - clim(t + \Delta t) \tag{4}$$

where $clim(t)$ and $clim(t + \Delta t)$ are climatology forecasts obtained from Equation (2) at t and $t + \Delta t$, respectively. The Equations (3) and (4) yield the forecast at $t + \Delta t$ as

$$Q(t + \Delta t) = Q(t) - clim(t) + clim(t + \Delta t) \tag{5}$$

The forecasts computed in Equations (1)–(5) use the entire streamflow time-series in record, i.e., at least 10 years of streamflow data. Here, we are also interested in exploring the forecast skill associated with the direct runoff component of streamflow. In other words, we computed forecast skill from persistence-based approaches for the rainfall-runoff events. Acknowledging the fact that the separation of runoff components from baseflow is not easy, we used the automated separation technique used by the United States Geological Survey (USGS) called the hydrograph separation program (HYSEP) [34]. The HYSEP employs three methods to separate storm flow from baseflow in the hydrograph: fixed interval, sliding interval, and local minimum. The duration of the surface runoff is computed empirically by

$$K = A^{0.2} \tag{6}$$

where K is the number of days after which the surface runoff stops and A is the upstream drainage area of the basin in sq. miles [34]. The interval used to separate the storm flow is $2K^*$, which should be the nearest odd integer to $2K$. The hydrograph separation starts one interval, i.e., $2K^*$ days, before the start of the date selected for the start of the separation, while it ends $2K^*$ days after the selected date. For further details of the separation technique, refer to Sloto and Crouse [34]. Figure A2 of the Appendix A shows the demonstration of storm flow separation across two river basins (small and large scales) using the sliding interval method.

2.3. Evaluation Metrics

We used several standard hydrologic forecast evaluation measures to evaluate the forecasts described in Section 2.2. The metrics we computed were Kling–Gupta efficiency (KGE), mean absolute error (MAE), normalized MAE ($nMAE$), timing of the hydrographs (T_H) and peak difference (PD). The KGE comprises three components: Pearson's correlation coefficient (r), variance ratio (α) and mean ratio (β) (refer to [35]). It is given by

$$KGE = 1 - \sqrt{(r-1)^2 + (\alpha-1)^2 + (\beta-1)^2} \tag{7}$$

where $\alpha = \frac{\sigma_f}{\sigma_o}$ *and* $\beta = \frac{\mu_f}{\mu_o}$. σ_f and σ_o correspond to the standard deviations of forecasts and observed streamflow, respectively. μ_f and μ_o correspond to the means of forecasts and observed streamflow, respectively. The ideal value of KGE is equal to 1, which can be obtained when all three components attain values of 1. For example, if the values of α and β are close to 1, the corresponding KGE is dominated by the correlation component. By construction, the KGE associated with the persistence-based forecasts is dominated by correlation. When the mean of the observations is used as a benchmark, the frontier of the KGE separating between the good model and the bad model is equal to −0.41 see [36]. In our discussion, we primarily focus on the KGE.

We compute the MAE associated with the forecasts as below:

$$MAE = \frac{\Sigma|Q_f - Q_o|}{N} \tag{8}$$

where Q_f is the streamflow forecast, Q_o is the observation and N is the number of data pairs. For fair comparison across the basin scales, we normalize MAE by the upstream drainage area to compute $nMAE$. The hydrograph timing (T_H) is computed as the number of hours a hydrograph requires to be shifted so that the cross-correlation between the forecasts and the observations is maximized. By construction, T_H is close to the lead time of the forecast. The peak difference (PD) in percentage is computed as

$$PD = \frac{peak_f - peak_o}{N} \times 100 \tag{9}$$

where $peak_f$ and $peak_o$ are peaks of the forecast and the observations, respectively.

3. Results

In this section, we present key results from our experiment evaluating persistence-based reference forecasts at 65 stream gauge stations in the Himalayan region in Nepal. First, we present results for three major river systems (basin-wise forecasts), and then for the entire region (regional forecasts).

3.1. Basin-Wise Forecasts

Our notion behind exploring basin-wise forecast skill is due to the variability of snow/glacier cover, storage conditions and hydroclimatic conditions across three major river basins. In Figure 2, we show the variability of the KGE metric with drainage area for three major river systems of Nepal (Karnali, Narayani and Koshi) across forecast lead times of 1, 3, 6 and 12-days. The result corresponds to the simple persistence forecasts. For simple persistence forecasts, the KGE is dominated by the correlation component r. For all three basins across most of the stations, KGE > 0.8 for the lead time of 1-day. The stronger relationship between the KGE and basin size emerges with the increasing forecast lead times. Note that the pattern is not as strong in the Koshi river basin, but the KGE values are relatively higher across basin scales. In addition, there are not many smaller monitored basins in the Koshi river system to exactly decide on the pattern relative to other two river systems. The overall results across three river systems clearly show a strong spatio-temporal dependence of simple persistence-based forecast skill.

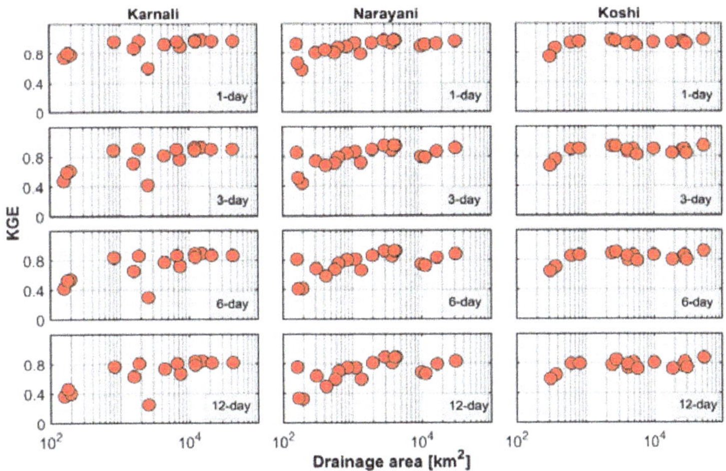

Figure 2. Relationship between the Kling–Gupta Efficiency (KGE) and drainage area across three major basins in Nepal. The columns show the Karnali, Narayani and Koshi river basins while the rows show lead times of 1, 3, 6 and 12 days, respectively.

3.2. *Regional Streamflow Forecasts*

The results from basin-wise forecasts show that no distinct pattern of forecast skill emerges across major river systems in Nepal. Therefore, we decided to pool all stations from three major river systems, including the mid-sized river stations from other river basins (see Figure 1), and refer to their forecasts as regional forecasts. In Figure 3, we present results in terms of two skill metrics, i.e., KGE (first row) and *nMAE* (second row), based on simple persistence forecasts (Equation (1)). The relationship between the KGE and basin size is more prominent for pooled stations across lead times of the forecast. There are, however, few stations which diverge from the overall pattern of the KGE observed in the basin-wise forecast evaluation. We explore the forecast performance associated with these stations separately (see Figure A3, Appendix A). We find that their forecast skills show more variability and decay much faster with forecast lead times. In addition, these stations are mostly mid-sized river basins associated with the intermittent flow regime. We present the detailed discussion on the origin of their forecast skill later. Note that basins > 1000 km^2 show relatively higher forecast skill (KGE~0.7) even at the lead time of 12 days. The values of *nMAE* also depict similar dependence with basin size as the KGE. The values of *nMAE* show a much stronger relationship with basin scale at shorter lead times, while starting to diverge at longer lead times particularly for small basin scales of size < 1000 km^2. By construction, *PD* for the simple persistence forecast is close to 0. In addition, by construction, *PD* for the climatology-based forecasts does not vary with lead times (see Figure A4, Appendix A). The corresponding median value of *PD* across basins in Nepal is about −76% (underestimation of the peak). For anomaly persistence, however, the median values of *PD* across basins range from about −1.5% to −3% at lead times of 1 day to 12 days, respectively (see Figure A4, Appendix A). Note that climatology-based forecasts show sizable a dependence of *PD* on basin size, while this is not apparent for anomaly persistence-based forecasts.

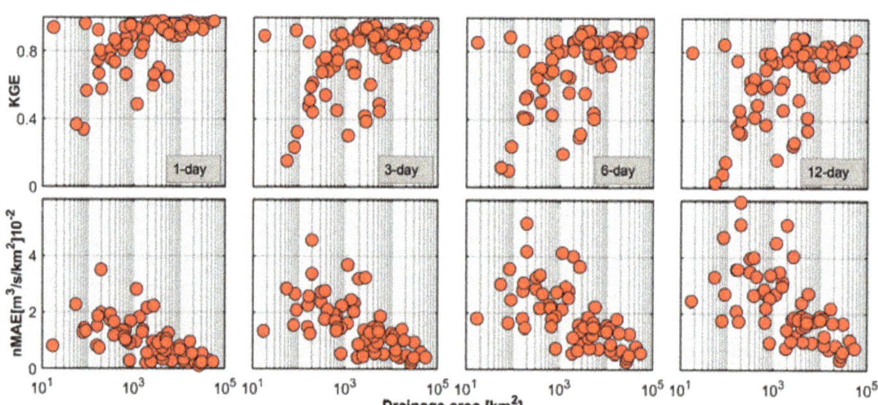

Figure 3. Relationship of skill metrics with drainage area for the entire Nepal. The columns show 1, 3, 6 and 12 day lead times, while the rows show Kling–Gupta efficiency (KGE) and normalized mean absolute error (*nMAE*), respectively.

In this section, we evaluate the forecast performance associated with various reference forecast schemes described in Section 2. Here, we present the one-to-one comparison of forecast skill in terms of KGE (first row) and *nMAE* (second row) between these schemes in Figure 4. Apparently, persistence-based forecasts outperform climatology-based forecasts at shorter lead times across all basin scales. Both simple persistence and climatology-based forecasts perform similarly for longer lead times. Anomaly persistence, however, performs similarly with simple persistence for shorter lead times, while it outperforms simple persistence for longer lead times at smaller basin scales (also see Figure 5). Note that climatology is an integral component of anomaly persistence forecasts. Therefore, the improved skill of anomaly persistence at longer lead times and smaller basin scales could be attributed to the improved performance of climatology-based forecasts at longer lead times.

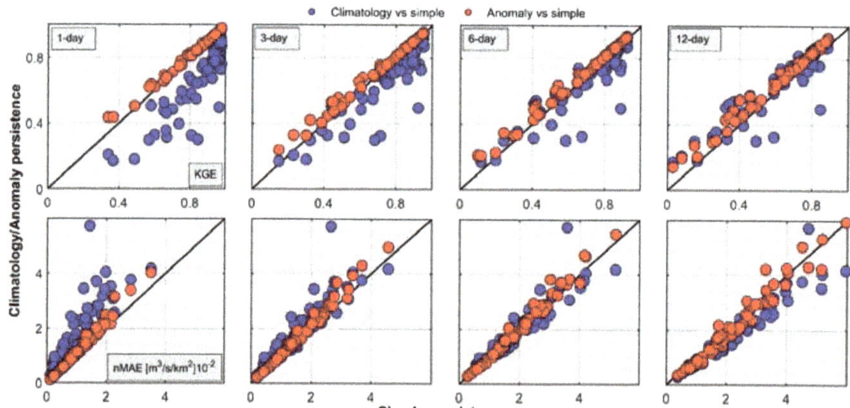

Figure 4. One-to-one relationship of skill metrics between climatology and simple persistence-based forecasts (blue), and between anomaly persistence-based forecasts and simple persistence-based forecasts (red). The columns show 1, 3, 6 and 12 day lead times, while the rows show Kling–Gupta efficiency (KGE) and normalized mean absolute error (*nMAE*), respectively.

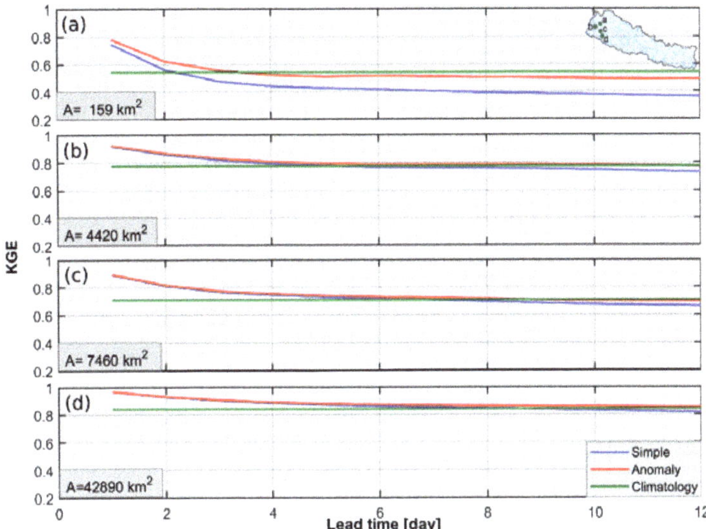

Figure 5. Evolution of KGE metric with lead time across increasing basin scales (**a–d**) shown in the inset. The panels show the KGE comparison between three methods: simple persistence, anomaly persistence and climatology.

From the real-time streamflow forecasting point of view, it would be more informative to explore how KGE evolves over time across basin scales. In Figure 5, we show that the evolution of KGE skill is associated with three methods across four nested basins in the Karnali river system (refer inset). In addition to the apparent strong dependence of KGE with basin scales, the temporal evolutive pattern of KGE shows variability across basin scales. For example, the smallest basin (A~159 km^2) depicts the sharp drop in the KGE metric up to lead time of four days and remains stable from thereon. However, as the basin size increases, the rate of decrease of the KGE metric becomes slower. The other important observation is that up to about four days of lead time, both persistence-based forecast schemes outperform climatology-based forecasts. After four days of lead time, the difference between climatology and anomaly persistence-based forecast skill is negligible. Note, however, that the difference in KGE between simple persistence and other two forecast schemes shows systematic decrease with the increasing basin scales.

The results presented above are associated with the entire streamflow time-series. In other words, the forecast skills computed originate from the contribution of both baseflow and stormflow. However, the intrinsic question is whether we can achieve similar forecast performance for the storm flow (i.e., direct runoff component of hydrographs). In Figure 6, we present the forecast performance based on simple persistence for the direct runoff obtained using HYSEP described in Section 2. As we demonstrate through stormflow hydrographs in Figure A5, Appendix A, at a lead time of one day, the simple persistence-based forecasts are essentially delayed by one day. In other words, the timing of forecast hydrographs and the timing of the extreme event peak are delayed by one day, while preserving the magnitude of the peak of the hydrograph. Note the damping of smaller rainfall-runoff event stormflows in the smaller nested basin (see a, Figure A5, Appendix A) by the river network aggregation process at the larger basin (see b, Figure A5, Appendix A). The resulting pattern of both KGE and $nMAE$ with basin scales is similar to the one presented before. Notably, reasonable KGE could be achieved for a one-day-ahead stormflow forecast, while showing significant decline for longer lead times. Note that the basins of size > 1000 km^2 still show KGE > 0.3 for 12 days ahead forecast, illustrating the potential of using it as reference forecast for the evaluation of stormflow forecasts particularly at short-range.

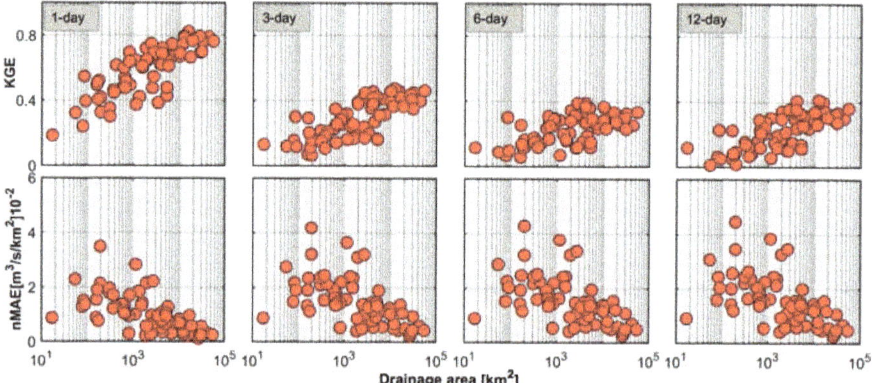

Figure 6. Relationship of skill metrics with drainage area for event-based streamflow. The columns show 1, 3, 6 and 12 day lead times, while the rows show Kling–Gupta efficiency (KGE) and normalized mean absolute error ($nMAE$), respectively.

As we highlighted in Section 2, our region predominantly contains rivers in the perennial flow regime (snow/glacier fed). There are, however, some rivers originating in the southern plains of Nepal which are predominantly in the intermittent flow regime. Since we established the fact that forecast skill shows strong spatial scale dependence, it would be of interest to the forecasting community to explore the dependence on the flow regimes. The ideal comparison would be to evaluate forecasts conditional on the basin scale. We identified four basins of a size of ~2500 km^2; two perennial river basins in northern Nepal and the other two intermittent river basins in southern Nepal (see inset of Figure 7). Figure 7 demonstrates the evolution of KGE (simple persistence-based) with lead times for these four basins. A distinct pattern of KGE evolution emerges. The perennial rivers depict higher forecast performance with gradual decline in the forecast performance with forecast horizon (lead time) while the intermittent rivers show sharp decline in the forecast performance with forecast horizon. Since intermittent rivers are mostly driven by rainfall-runoff events, their forecast skill evolution shows similar behavior elucidated by the results in Figure 6. Note a sudden spike of KGE for the Babai river at four days of lead time. Though it is hard to point out explicitly the reason, one could attribute it to the ability of four days ahead forecast to capture two consecutive historic floods (separated by 4 days) in the year 2015.

It is extremely important for forecasters to achieve good predictability of higher flow quantiles for flood preparedness and mitigation efforts. For the same four basins, we evaluate the forecast performance associated with the different streamflow quantiles. We refer to flow quantiles as flow threshold. Figure 8 shows the evolution of KGE across lead times for various flow quantiles. KGE reported here is computed for the streamflow forecasts exceeding the corresponding flow quantile. Note that as the flow quantile increases, the sample size used to compute the forecast skill systematically decreases. Given that we evaluate the forecast skill using the continuous streamflow time-series in record, we consider the sample size to be enough for the evaluation. Clearly, the KGE for the intermittent rivers (lower panel) shows sharp decline for the higher flow quantiles. The perennial rivers (upper panel), however, depict much better forecast performance even for the higher flow quantiles at longer lead times. However, of the two, the Marshyangdi basin at Bhakundebesi, shows relatively faster decline in the performance of flows exceeding 60th percentile. As Ghimire and Krajewski [19] highlighted, this variability in KGE could be explained by the difference in their river network geometries, typically explained by the network width function see [37,38] for detail. Explicit attribution is beyond the scope of this paper.

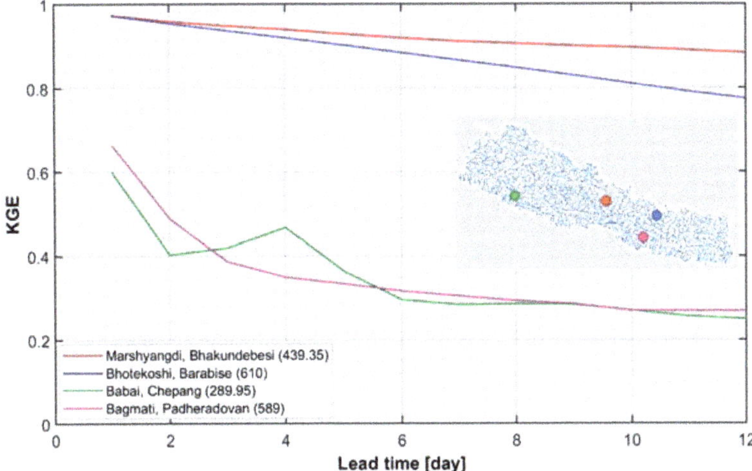

Figure 7. Evolution of Kling-Gupta efficiency (KGE) metric with lead time. The lines in the top (red and blue) represent snow-fed river basins while the lines in the bottom (green and magenta) represent intermittent river basins. Note that all four river basins are of a similar size of ~2500 km^2.

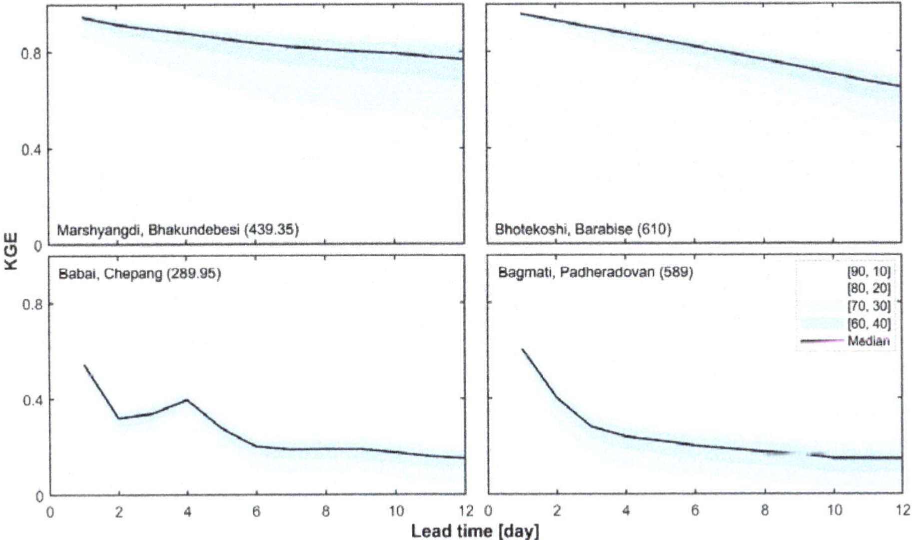

Figure 8. Percentile plots showing the evolution of Kling–Gupta efficiency (KGE) with lead time for four river basins described in Figure 7. Each percentile refers to the KGE computed for streamflow forecasts exceeding corresponding percentile flow. For example, 90th percentile KGE represents the KGE computed for streamflow forecasts >90th percentile flow. The envelope corresponds to each 10th quantile on either side of the median flow. The higher and lower flow quantiles shown in the interval correspond to the lower and upper bounds of the envelope, respectively. For example, the interval [90, 10] corresponds to the KGE computed for 90th and 10th percentile flows, respectively.

4. Discussion

A clear signature emerging out of our persistence-based streamflow forecasts skill is its strong spatio-temporal dependence. Our results substantiate it further. The results from Figures 3 and 6 demonstrate that significant contribution to the persistence-based forecast skill is from the baseflow contribution. In other words, the contribution from snowpack (snow/glacier), groundwater and storage in the contributing catchments plays an important role in the streamflow hydrographs, particularly for the perennial rivers in the Himalayan region of Nepal. Typically, the streamflow predictability at small basin scales is more tied to the rainfall than the large basin scales. We highlighted this fact through our results for the intermittent rivers. The perennial rivers, however, receive sizable contributions from the base flow relative to the rainfall even in the small basin scales, hence the relatively good forecast skill. Moreover, the persistence-based forecast derives skill from the memory of the system, in this case river catchments. The larger the basin scales, the longer the memory of the catchments, and hence the long-range persistence. We show from our results that higher forecast skill (KGE > 0.7) could be achieved for a basin size larger than 1000 km^2, even at a longer forecast horizon. Another important contribution to the skillful forecasts of larger basins at longer lead times is from the water transport in the river network, where the streamflow aggregation process controls the shape of the streamflow fluctuations [27,39–41]. In addition to the land surface (e.g., catchment) memory, the streamflow predictability has a strong connection to the persistence in the land surface initial hydrologic conditions (e.g., soil moisture, groundwater, current streamflow and snowpack) [42–44], providing the main source for the skillful streamflow forecasts.

The dependence of forecast skill depicted in this study with the basin size and forecast lead times is consistent with the results presented in Ghimire and Krajewski [19]. Given a large number of glacierized basins in the region, as opposed to agricultural watersheds used in the study of Ghimire and Krajewski [19], more stations depict higher streamflow forecast skill. For instance, at a lead time of one day, we show a median value of KGE of about 0.92 in the region, compared to the median value of KGE of about 0.78 presented in their study. Therefore, our results demonstrate a clear implication of persistence-based forecasts for the real-time streamflow forecasting in the Himalayan region. Our illustration of reasonably good streamflow forecast skill across spatial and temporal scales provides a reliable benchmark (reference) to evaluate the efficacy of the operational real-time streamflow forecasting. Particularly for the larger basin scales, we are able to show that it is difficult for many operational mechanistic hydrologic models to outperform persistence-based forecast skills. Any operational forecasting scheme that can aptly depict forecast skill better than the three-reference forecast schemes presented in this study, can be considered skillful. We are able to show quantitatively that it can serve as a skillful reference for the evaluation of real-time flood forecasts (higher flow quantiles) up to the medium range for perennial rivers while up to the short-range for the intermittent rivers. Our findings clearly show that persistence-based forecast scheme, anomaly persistence in particular, could provide skillful reference forecasts for the evaluation of current operational streamflow forecasting systems in the Himalayan country of Nepal. Moreover, it could provide guidance to the flood related decision-making process, especially in the short-medium forecast range.

5. Conclusions

In this study, we explored the utility of persistence-based forecasting schemes to benchmark the real-time streamflow forecasting system in the Himalayan region of Nepal. We used the daily streamflow data at 65 stream gauge stations monitored by the DHM, Nepal, to generate the forecasts and evaluate the associated skills in the short-to-medium forecast horizon. To this end, we employed three reference forecast schemes: (i) simple persistence, (ii) streamflow climatology and (iii) anomaly persistence. Based on the results from this study, we could reach at following conclusions:

- Persistence-based forecast skill shows strong dependence with the basin scale and forecast lead time. Anomaly persistence forecasts outperform others at small basin scales and longer lead times and hence can be a better selection for benchmarking the real-time streamflow forecasting system in the Himalayan region of Nepal.

- The forecast skill shows strong dependence with the flow regime and flow threshold. The verification results show higher forecast skill for perennial rivers over intermittent rivers and moderate flow quantiles over high flow quantiles.

This study highlights the fact that a persistence-based forecast scheme is difficult to outperform using many mechanistic hydrologic models, particularly for larger basin scales in the Himalayan region. The findings from this study have implications for evaluating real-time streamflow forecasts from the operational forecasting system across basin scales and lead times in this region. Moreover, it provides insights on designing the streamflow monitoring network for future applications.

Our study, however, is not without limitations. We did not account for the measurement uncertainty associated with streamflow observations. We considered the published streamflow data to be within the acceptable limit of observational uncertainty. We did not account for the uncertainty associated with the rating curves. Moreover, we could not perform the sub-daily forecast skill evaluation due to the unavailability of published sub-daily streamflow observations. We expect that the analysis using sub-daily streamflow data will not lead to a significantly different inference regarding the forecasting performance of persistence-based systems, which we could explore further in our future research.

Author Contributions: Conceptualization, G.R.G.; methodology, G.R.G., and S.S.; software, G.R.G.; formal analysis, G.R.G.; investigation, G.R.G., S.S., J.P., R.T., B.P., P.D., and R.B.; resources, G.R.G., B.P., J.P.; writing—original draft preparation, G.R.G.; writing—review and editing, G.R.G., S.S., J.P., R.T., B.P., P.D., and R.B. All authors have read and agreed to the published version of the manuscript.

Funding: This research received no external funding.

Acknowledgments: The authors greatly acknowledge the Department of Hydrology and Meteorology (DHM), Nepal for providing the streamflow data used for the study. The authors are grateful to the anonymous reviewers for their reviews and constructive comments.

Conflicts of Interest: The authors declare no conflict of interest.

Appendix A

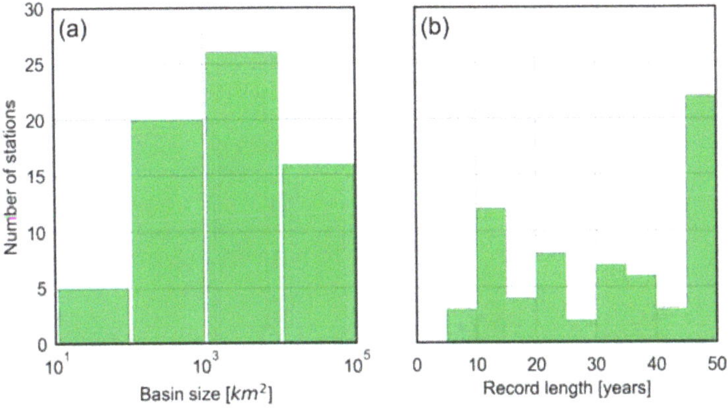

Figure A1. Histograms showing the distribution of upstream drainage area (**a**) and streamflow record length (**b**) associated with the stream gauge stations in Nepal.

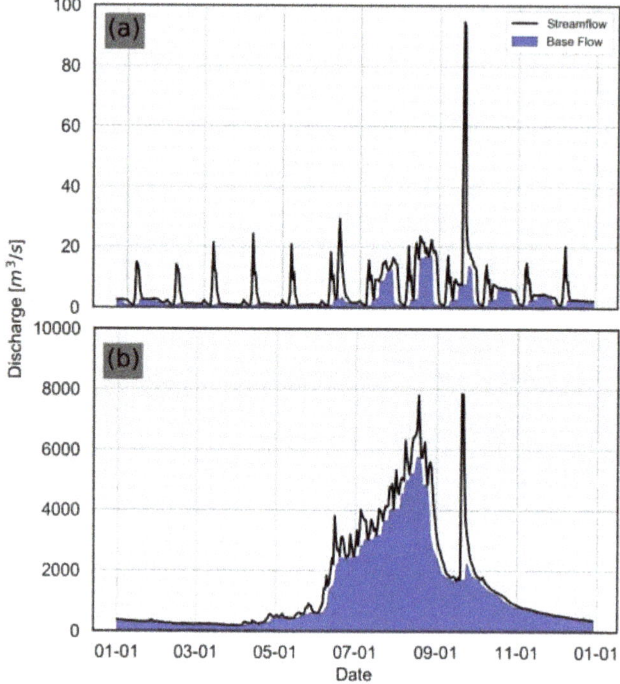

Figure A2. Demonstration of streamflow separation using HYSEP. Area highlighted in the blue represents the baseflow while the remaining area represents the direct run-off component. (**a**,**b**) correspond to the Langurkhola in Chhana (159 km^2) and Karnali river in Chisapani (42,890 km^2), respectively.

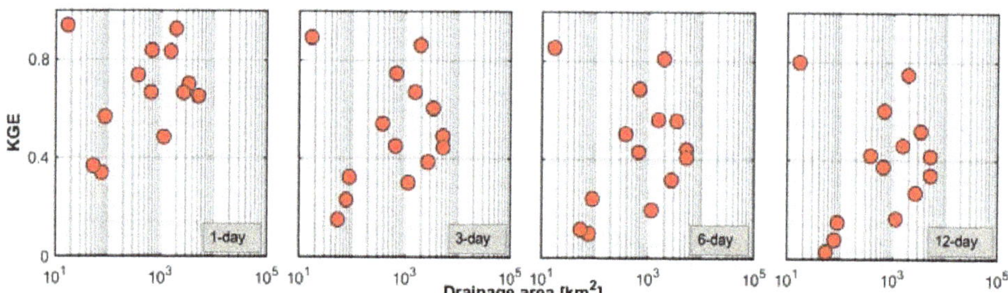

Figure A3. Relationship between KGE and drainage area across lead times for the intermittent flow rivers.

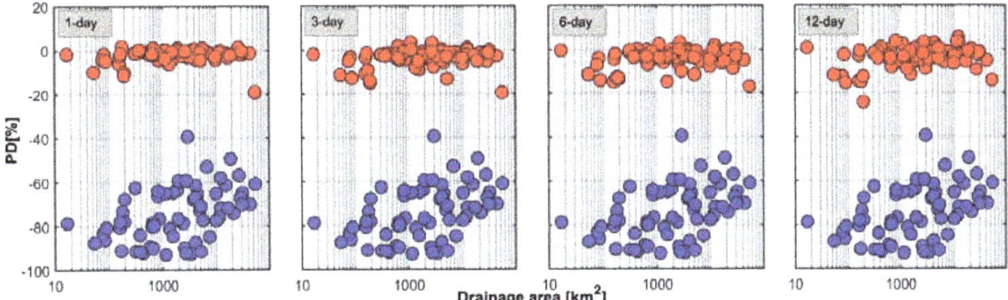

Figure A4. Relationship between peak difference, *PD* and drainage area across lead times. Blue dots correspond to the climatology forecasts and red dots correspond to the anomaly persistence-based forecasts.

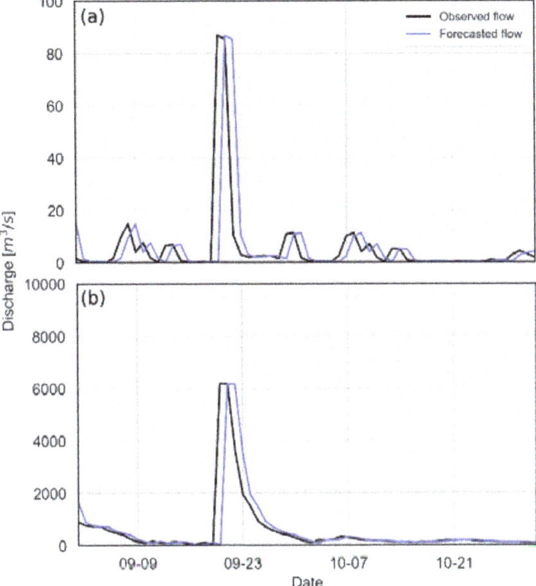

Figure A5. Demonstration of observed and forecasted stormflow hydrographs for the largest rainfall-runoff event of the year 2008 between 9/1/2008 and 11/1/2008 based on simple persistence for two basins shown in Figure A2 at a lead time of one day: Langurkhola in Chhana (159 km^2) and Karnali river in Chisapani (42,890 km^2), respectively.

References

1. Palash, W.; Jiang, Y.; Akanda, A.S.; Small, D.L.; Nozari, A.; Islam, S. A Streamflow and Water Level Forecasting Model for the Ganges, Brahmaputra, and Meghna Rivers with Requisite Simplicity. *J. Hydrometeorol.* **2018**, *19*, 201–225. [CrossRef]
2. Pellicciotti, F.; Buergi, C.; Immerzeel, W.W.; Konz, M.; Shrestha, A.B. Challenges and Uncertainties in Hydrological Modeling of Remote Hindu Kush-Karakoram-Himalayan (HKH) Basins: Suggestions for Calibration Strategies. *Mt. Res. Dev.* **2012**, *32*, 39–50. [CrossRef]
3. Sharma, S.; Siddique, R.; Reed, S.; Ahnert, P.; Mendoza, P.; Mejia, A. Relative effects of statistical preprocessing and postprocessing on a regional hydrological ensemble prediction system. *Hydrol. Earth Syst. Sci.* **2018**, *22*, 1831–1849. [CrossRef]

4. Gurung, D.R.; Giriraj, A.; Aung, K.S.; Shrestha, B.; Kulkarni, A.V. *Snow-Cover Mapping and Monitoring in the Hindu Kush-Himalayas*; The International Centre for Integrated Mountain Development: Lalitpur, Nepal, 2011; pp. 1–44.
5. Khadka, D.; Babel, M.S.; Shrestha, S.; Tripathi, N.K. Climate change impact on glacier and snow melt and runoff in Tamakoshi basin in the Hindu Kush Himalayan (HKH) region. *J. Hydrol.* **2014**, *511*, 49–60. [CrossRef]
6. Paudel, K.P.; Andersen, P. Monitoring snow cover variability in an agropastoral area in the Trans Himalayan region of Nepal using MODIS data with improved cloud removal methodology. *Remote Sens. Environ.* **2011**, *115*, 1234–1246. [CrossRef]
7. Mukherji, A.; Molden, D.; Nepal, S.; Rasul, G.; Wagnon, P. Himalayan waters at the crossroads: Issues and challenges. *Int. J. Water Resour. Dev.* **2015**, *31*, 151–160. [CrossRef]
8. Bhattarai, B.C.; Regmi, D. Impact of Climate Change on Water Resources in View of Contribution of Runoff Components in Stream Flow: A Case Study from Langtang Basin, Nepal. *J. Hydrol. Meteorol.* **2016**, *9*, 74–84. [CrossRef]
9. Dhami, B.; Himanshu, S.K.; Pandey, A.; Gautam, A.K. Evaluation of the SWAT model for water balance study of a mountainous snowfed river basin of Nepal. *Environ. Earth Sci.* **2018**, *77*, 1–20. [CrossRef]
10. Government of Nepal Ministry of Energy Water Resources and Irrigation; Department of Hydrology and Meteorology. *Standard Operating Procedure for Flood Early Warning System in Nepal*; Department of Hydrology and Meteorology: Kathmandu, Nepal, 2018.
11. Gautam, D.K.; Phaiju, A.G. Community Based Approach to Flood Early Warning in West Rapti River Basin of Nepal. *J. Integr. Disaster Risk Manag.* **2013**, *3*, 155–169. [CrossRef]
12. Smith, P.J.; Brown, S.; Dugar, S. Community-based early warning systems for flood risk mitigation in Nepal. *Nat. Hazards Earth Syst. Sci.* **2017**, *17*, 423–437. [CrossRef]
13. Budimir, M.; Donovan, A.; Brown, S.; Shakya, P.; Gautam, D.; Uprety, M.; Cranston, M.; Sneddon, A.; Smith, P.; Dugar, S. Communicating Complex Forecasts for Enhanced Early Warning in Nepal. *Geosci. Commun. Discuss.* **2019**, 1–32. [CrossRef]
14. Liu, Y.; Gupta, H.V. Uncertainty in hydrologic modeling: Toward an integrated data assimilation framework. *Water Resour. Res.* **2007**, *43*, 1–18. [CrossRef]
15. Sharma, S.; Siddique, R.; Balderas, N.; Fuentes, J.D.; Reed, S.; Ahnert, P.; Shedd, R.; Astifan, B.; Cabrera, R.; Laing, A.; et al. Eastern U.S. verification of ensemble precipitation forecasts. *Weather Forecast.* **2017**, *32*, 117–139. [CrossRef]
16. Pappenberger, F.; Scipal, K.; Buizza, R. Hydrological aspects of meteorological verification. *Atmos. Sci. Lett.* **2008**, *9*, 43–52. [CrossRef]
17. Kayastha, R.B.; Steiner, N.; Kayastha, R.; Mishra, S.K.; Forster, R.R. Comparative Study of Hydrology and Icemelt in Three Nepal River Basins Using the Glacio-Hydrological Degree-Day Model (GDM) and Observations From the Advanced Scatterometer. *Front. Earth Sci.* **2020**, *7*, 1–13. [CrossRef]
18. WRI Aqueduct Global Flood Risk Country Rankings. 2015. Available online: https://www.wri.org/resources/data-sets/aqueduct-global-flood-risk-country-rankings (accessed on 29 June 2020).
19. Ghimire, G.R.; Krajewski, W.F. Exploring Persistence in Streamflow Forecasting. *J. Am. Water Resour. Assoc.* **2020**, *56*, 542–550. [CrossRef]
20. Mittermaier, M.P. The Potential Impact of Using Persistence as a Reference Forecast on Perceived Forecast Skill. *Weather Forecast.* **2008**, *23*, 1022–1031. [CrossRef]
21. Bennett, J.C.; Robertson, D.E.; Shrestha, D.L.; Wang, Q.J. Selecting reference streamflow forecasts to demonstrate the performance of NWP-forced streamflow forecasts. In Proceedings of the 20th International Congress on Modelling and Simulation, Adelaide, Australia, 1–6 December 2013; pp. 2611–2617.
22. Ghimire, G.R.; Jadidoleslam, N.; Krajewski, W.F.; Tsonis, A.A. Insights On Streamflow Predictability Across Scales Using Horizontal Visibility Graph Based Networks. *Front. Water* **2020**. accepted for publication.
23. van den Dool, H. *Empirical Methods in Short-Term Climate Prediction*; Oxford University Press: Oxford, UK, 2007; ISBN 0-19-920278-8/978-0-19-920278-2.
24. Fraedrich, K.; Ziehmann-Schlumbohm, C. *Predictability Experiments with Persistence Forecasts in a Red-noise Atmosphere*; Royal Meteorological Society: Reading, UK, 1994; ISBN 0035-9009.
25. Wu, W.; Dickinson, R.E. Warm-season rainfall variability over the U.S. Great Plains and its correlation with evapotranspiration in a climate simulation. *Geophys. Res. Lett.* **2005**, *32*, 215. [CrossRef]

26. Pagano, T.; Garen, D. A Recent Increase in Western U.S. Streamflow Variability and Persistence. *J. Hydrometeorol.* **2005**, *6*, 173–179. [CrossRef]
27. Krajewski, W.F.; Ghimire, G.R.; Quintero, F. Streamflow Forecasting without Models. *J. Hydrometeorol.* **2020**, in press.
28. Karki, R.; Talchabhadel, R.; Aalto, J.; Baidya, S.K. New climatic classification of Nepal. *Theor. Appl. Climatol.* **2016**, *125*, 799–808. [CrossRef]
29. Uddin, K.; Shrestha, H.L.; Murthy, M.S.R.; Bajracharya, B.; Shrestha, B.; Gilani, H.; Pradhan, S.; Dangol, B. Development of 2010 national land cover database for the Nepal. *J. Environ. Manag.* **2015**, *148*, 82–90. [CrossRef]
30. Shrestha, M.L. Interannual variation of summer monsoon rainfall over Nepal and its relation to Southern Oscillation Index. *Meteorol. Atmos. Phys.* **2000**, *75*, 21–28. [CrossRef]
31. Mool, P.K.; Wangda, D.; Bajracharya, S.R.; Joshi, S.P.; Kunzang, K.; Gurung, D.R. *Inventory of Glaciers, Glacial Lakes and Glacial Lake Outburst Floods: Monitoring and Early Warning Systems in the Hindu Kush-Himalayan Region—Bhutan*; International Centre for Integrated Mountain Development: Khumaltar, Nepal, 2001.
32. DHM, N. *Climate Normals*; The Department of Hydrology and Meteorology: Kathmandu, Nepal, 2010.
33. Gautam, M.R.; Acharya, K. Streamflow trends in Nepal. *Hydrol. Sci. J.* **2012**, *57*, 344–357. [CrossRef]
34. Sloto, R.a.; Crouse, M.Y. *Hysep: A Computer Program for Streamflow Hydrograph Separation and Analysis*; Water-Resources Investigations Report 96-4040; U.S. Geological Survey: Reston, VA, USA, 1996; p. 54.
35. Gupta, H.V.; Kling, H.; Yilmaz, K.K.; Martinez, G.F. Decomposition of the mean squared error and NSE performance criteria: Implications for improving hydrological modelling. *J. Hydrol.* **2009**, *377*, 80–91. [CrossRef]
36. Knoben, W.J.M.; Freer, J.E.; Woods, R.A. Technical note: Inherent benchmark or not? Comparing Nash-Sutcliffe and Kling-Gupta efficiency scores. *Hydrol. Earth Syst. Sci. Discuss.* **2019**, *23*, 4323–4331. [CrossRef]
37. Rodriguez-Iturbe, I.; Rinaldo, A. *Fractal River Basins: Chance and Self-Organization*; Cambridge University Press: Kolkata, India, 1997; ISBN 0521004055.
38. Perez, G.; Mantilla, R.; Krajewski, W.F. The influence of spatial variability of width functions on regional peak flow regressions. *Water Resour. Res.* **2018**, *54*, 7651–7669. [CrossRef]
39. Ayalew, T.B.; Krajewski, W.F.; Mantilla, R. Connecting the power-law scaling structure of peak-discharges to spatially variable rainfall and catchment physical properties. *Adv. Water Resour.* **2014**, *71*, 32–43. [CrossRef]
40. Ayalew, T.B.; Krajewski, W.F.; Mantilla, R.; Small, S.J. Exploring the effects of hillslope-channel link dynamics and excess rainfall properties on the scaling structure of peak-discharge. *Adv. Water Resour.* **2014**, *64*, 9–20. [CrossRef]
41. Ayalew, T.B.; Krajewski, W.F.; Mantilla, R. Analyzing the effects of excess rainfall properties on the scaling structure of peak discharges: Insights from a mesoscale river basin. *Water Resour. Res.* **2015**, *51*, 3900–3921. [CrossRef]
42. Arnal, L.; Wood, A.W.; Stephens, E.; Cloke, H.L.; Pappenberger, F. An efficient approach for estimating streamflow forecast skill elasticity. *J. Hydrometeorol.* **2017**, *18*, 1715–1729. [CrossRef]
43. Harrigan, S.; Prudhomme, C.; Parry, S.; Smith, K.; Tanguy, M. Benchmarking ensemble streamflow prediction skill in the UK. *Hydrol. Earth Syst. Sci.* **2018**, *22*, 2023–2039. [CrossRef]
44. Wood, A.W.; Pagano, T.; Roos, M. Tracing The Origins of ESP, HEPEX Blog. 2016. Available online: https://hepex.inrae.fr/tracing-the-origins-of-esp/ (accessed on 29 June 2020).

Article

A Generalized Flow for B2B Sales Predictive Modeling: An Azure Machine-Learning Approach

Alireza Rezazadeh

Electrical and Computer Engineering Department, University of Illinois at Chicago, Chicago, IL 60607, USA; arezaz2@uic.edu

Received: 3 July 2020; Accepted: 5 August 2020; Published: 6 August 2020

Abstract: Predicting the outcome of sales opportunities is a core part of successful business management. Conventionally, undertaking this prediction has relied mostly on subjective human evaluations in the process of sales decision-making. In this paper, we addressed the problem of forecasting the outcome of Business to Business (B2B) sales by proposing a thorough data-driven Machine-Learning (ML) workflow on a cloud-based computing platform: Microsoft Azure Machine-Learning Service (Azure ML). This workflow consists of two pipelines: (1) An ML pipeline to train probabilistic predictive models on the historical sales opportunities data. In this pipeline, data is enriched with an extensive feature enhancement step and then used to train an ensemble of ML classification models in parallel. (2) A prediction pipeline to use the trained ML model and infer the likelihood of winning new sales opportunities along with calculating optimal decision boundaries. The effectiveness of the proposed workflow was evaluated on a real sales dataset of a major global B2B consulting firm. Our results implied that decision-making based on the ML predictions is more accurate and brings a higher monetary value.

Keywords: costumer relation management; business to business sales prediction; machine learning; predictive modeling; microsoft azure machine-learning service

---

1. Introduction

In the Business to Business (B2B) commerce, companies compete to win high-valued sales opportunities to maximize their profitability. In this regard, a key factor for maintaining a successful B2B enterprise is the task of forecasting the outcome of sales opportunities. B2B sales process typically demands significant costs and resources and, hence, requires careful evaluations in the very early steps. Quantifying the likelihood of winning new sales opportunities is an important basis for appropriate resource allocation to avoid wasting resources and sustain the company's financial objectives [1–4].

Conventionally, forecasting the outcome of sales opportunities is carried out mostly relying on subjective human rating [5–8]. Most of the Customer Relationship Management (CRM) systems allow salespersons to manually assign a probability of winning for new sales opportunities [9]. This probability is then used at various stages of the sales pipeline, i.e., calculating a weighted revenue of the sales records [10,11]. Often each salesperson develops a non-systematic intuition to forecast the likelihood of winning a sales opportunity with little to no quantitative rationale, neglecting the complexity of the business dynamics [9]. Moreover, as often as not, sales opportunities are intentionally underrated to avoid any internal competition with other salespersons or overrated to circumvent the pressure from sales management to maintain a higher performance [12].

Even though the abundance of data and improvements in statistical and machine-learning (ML) techniques have led to significant enhancements in data-driven decision-making, the literature is scarce in the subject of B2B sales outcome forecasting. Yan et al. [12] explored predicting win-propensity of sales opportunities using a two-dimensional Hawkes dynamic clustering technique. Their approach allowed for live assessment of active sales although relied heavily on regular updates and inputs from salespersons in the CRM system. This solution is hard to maintain in larger B2B firms considering each salesperson often handles multiple opportunities in parallel and would put less effort into making frequent interaction with each sales record [13].

Tang et al. [9] built a sales forecast engine consist of multiple models trained on snapshots of historical data. Although their paradigm is focused on revenue forecasting, they demonstrated the effectiveness of hybrid models for sales predictive modeling. Bohane et al. [5] explored the idea of single and double-loop learning in B2B forecasting using ML models coupled with general explanation methods. Their main goal was actively involving users in the process of model development and testing. Built on their earlier work on effective feature selection [14] they concluded random forest models were the most promising for B2B sales forecasting.

Here, we proposed an end-to-end cloud-based workflow to forecast the outcome of B2B sales opportunities by reframing this problem into a binary classification framework. First, an ML pipeline extracts sales data and improves them through a comprehensive feature enhancement step. The ML pipeline optimally parameterizes a hybrid of probabilistic ML classification models trained on the enhanced sales data and eventually outputs a voting ensemble classifier. Second, a prediction pipeline makes use of the optimal ML model to forecast the likelihood of winning new sales opportunities. Importantly, the prediction pipeline also performs thorough statistical analysis on the historical sales data and specifies appropriate decision boundaries based on sales monetary value and industry segment. This helps to maximize the reliability of predictions by binding the interpretation of model results to the actual data.

The proposed workflow was implemented and deployed to a global B2B consulting firm's sales pipeline using Microsoft Azure Machine-Learning Service (Azure ML). Such a cloud-based solution readily integrates into the existing CRM systems within each enterprise and allows for more scalability. Finally, we compared the performance of the proposed solution with salespersons' predictions using standard statistical metrics (e.g., accuracy, AUC, etc.). To make the comparison more concrete, we also investigated the financial aspect of implementing this solution and compared the monetary value of our ML solution with salespersons' predictions. Overall, we have found that the proposed ML solution results in a superior prediction both in terms of statistical and financial evaluations; therefore, it would be a constructive complement to the predictions made by salespersons.

This paper is organized as follows: In Section 2, materials and methods used in this work are introduced in detail. Section 3 summarizes the results of this work. Section 4 presents discussion on the results, limitations of the current work and potential future directions.

2. Materials and Methods

2.1. Data and Features

Data for this study were obtained from a global multi-business B2B consulting firm's CRM database in three main business segments: Healthcare, Energy, and Financial Services (Finance for short). This section, first, gives an overview of the data and then explains a data enhancement technique used to infer additional relevant information from the dataset.

Data

A total number of 25,578 closed sales opportunity records starting January 2015 through August 2019 were used in this work (Figure 1a). Each closed opportunity record contained a status label (won/lost) corresponding to its ultimate outcome, otherwise if still active in the sales pipeline, they were labeled as open. Out of all closed sales records $\sim$58% were labeled as "won" in their final sales status (Figure 1b).

A total number of 20 relevant variables (features) were extracted for each sales opportunity from the raw CRM database. Table 1 describes these features in more details. Specifically, a subset of the features described the sales project (Opportunity Type, General Nature of Work, Detailed Nature of Work, Project Location, Project Duration, and Total Contract Value, Status). The remaining features provided further information on the customer (Account, Account Location, Key Account Energy, Key Account Finance, and Key Account Healthcare) and the internal project segmentation and resource allocation (Business Unit, Engagement Manager, Sales Lead, Probability, Sub-practice, Practice, Group Practice, Segment, and User-entered Probability).

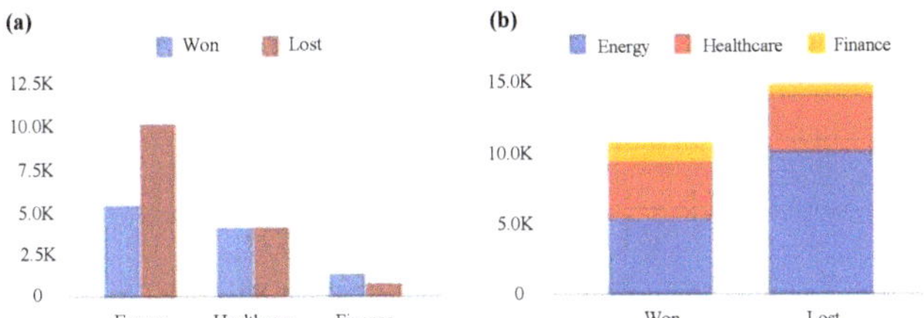

Figure 1. Data Exploration: (**a**) Distribution of sales opportunity records across the three business segments: Healthcare, Energy, and Finance. (**b**) Closed sales opportunities final status.

Table 1. Raw CRM sales database features.

Count	Feature Type	Features
13	Categorical	Business Unit, Opportunity Type, Project Location, General Nature of Work, Detailed Nature of Work, Account, Account Location, Sales Lead Engagement Manager, Sub-practice, Practice, Group Practice, Segment
4	Binary	Status, Key Account Energy, Key Account Healthcare, Key Account Finance
3	Continuous	User-entered Probability, Project Duration, Total Contract Value

Once a sales opportunity profile was created in the CRM system, users were required to input their estimation for the probability of winning that opportunity. Please note that the user-entered probabilities were not used in the process of training ML models and were only used as a point of reference to compare with the performance of the ML workflow. All the features listed in Table 1 were required to populate in the CRM system; therefore, less than 1% of the dataset contained missing values. As a result, sales records with a missing value were dropped from the dataset.

2.2. *Feature Enhancement*

The CRM raw dataset was enhanced by inferring additional relevant features calculated across the sales records. These additional features were calculated using statistical analysis on the categorical features: Sales Leads, Account, Account Location, etc. Mainly, the idea was to extract a lookup table containing relevant statistics calculated across the sales records for each of the unique values in the categorical features.

By collecting the historical data of unique values of each categorical features (i.e., for each individual Sales Lead, Account, and Project Location, etc.), we calculated the following statistical metrics: (1) Total number of sales opportunities (2) Total number of won sales (3) Total number of lost sales (4) Average contract value (value for short) of won sales (5) Standard error of the mean won sales value (6) Winning rate calculated as the ratio of won and total sales counts (7) Coefficient of variation (the ratio of the standard deviation to the mean) [15] of won sales value to capture the extent of variability in the won sales contract values.

The aforementioned statistics were calculated and stored in feature enhancement lookup tables for each categorical feature (see Table 1 for a list of these features). Table 2 provides an example of a feature enhancement lookup table calculated based on the "Sales Lead" feature in the raw CRM dataset. These lookup Tables (13 tables in total for all categorical features) were appropriately merged back to the raw CRM sales dataset.

In the last feature enhancement step, the Mahalanobis [16] distance was calculated between each sales opportunity's value and the distribution of all won sales value that shared a similar categorical feature (individually for each of the 13 categorical features). This quantifies how far a sales value is relative to the family of won sales with the same characteristics (i.e., same Sales Lead, Project Location, Segment, etc.). The process of feature enhancement increased the total number of features to 137 for each sales record (20 features originally from the raw CRM dataset + $9 \times 13 = 117$ additional features from the lookup tables).

Table 2. Feature Enhancement Lookup Table: An example of the statistics calculated based on "Sales Lead" including counts of the total, won, and lost opportunities along with the mean and standard error of the mean (SEM) for won sales value and their coefficient of variation (CV).

	Sales Lead	Total	Won	Lost	Won Value Mean	Won Value SEM	Win Rate	CV
1	John Doe 1	3788	1902	1866	107,249.3	15,460.3	0.5	6.9
2	John Doe 2	1908	1908	0	3793.6	38.9	1.0	97.5
3	John Doe 3	1335	1232	103	5352.0	218.1	0.9	24.5
4	John Doe 4	986	454	566	492,626.8	90,913.8	0.5	5.4
5	John Doe 5	973	359	614	15,700.3	1283.0	0.4	12.2
...	...	...	...	...	...	...	...	...

The enhanced CRM dataset (25,578 total number of sales opportunities) was randomly split into a "train set" (70%) and a "test set" (30%). The train set was used to train ML models. The performance of the model on train set is reported using a 10-fold cross-validation technique. The test set was used to report the performance of the trained ML model on the unseen portion of the dataset. For further evaluations, after the proposed framework was deployed to the sales pipeline of the enterprise, a "validation set" was collected of new sales records over a period of 3 months (846 closed sales opportunities).

2.3. *Machine-Learning Overview*

Our solution to predicting the likelihood of winning sales opportunities is essentially reframing this problem in a supervised binary classification paradigm (won, lost). Hence, we made use of two of the most promising supervised classification algorithms: XGBoost, and LightGBM. In particular, these two models were selected among the family of probabilistic classification models due to their higher classification accuracy in our problem. A second motivation for using these two models was the fact that the distributed versions of both can easily integrate into cloud platforms such as Azure ML. Last, to attain a superior performance, multiple iterations of both models were combined in a voting ensemble.

2.3.1. Binary Classification

Probabilistic classification algorithms [17], given pairs of samples and their corresponding class labels $(X_1, y_1), \ldots, (X_n, y_n)$, capture a conditional probability distribution over the output classes $P(y_i \in Y \mid X_i)$ where for a binary classification scenario $Y \in \{0, 1\}$ (maps to lost/won in our problem). Given the predicted probability of a data sample, a decision boundary is required to define a reference point and predict which class the sample belongs to. In a standard binary classification, the predicted class is the one that has the highest probability [18]. This translates to a standard decision boundary of 0.5 for predicting class labels.

However, the decision boundary can be calibrated arbitrarily to reflect more on the distribution of the data. The influence of the decision boundary on the number of true positives (TP), false positives (FP), true negatives (TN), and false negatives (FN) in binary classification is illustrated in Figure 2 (see

Table 3 for definitions). In this work, we find the optimal decision boundary for a classification model by maximizing all true conditions (both TP and TN) which in return, minimizes all the false conditions (FP and FN). Visually, this decision boundary is a vertical line passing through the intersections of $P(X|Y=0)$ and $P(X|Y=1)$ in Figure 2.

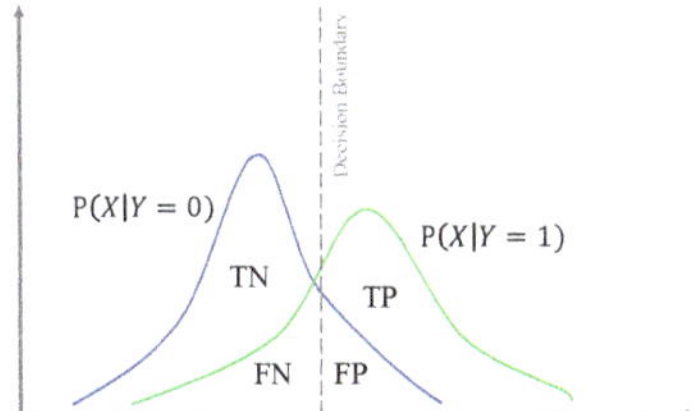

Figure 2. Decision Boundary: impact of the decision boundary on different scenarios of the binary classification.

The performance of a binary classifier can be evaluated using standard statistical metrics such as accuracy, precision, recall, and F1-score (see Table 3). For the case of binary classification, the area under ROC curve (AUC) measures the robustness of the classification (a higher AUC suggests more robust classification performance) [19]. As shown by Hand et al. [20], the AUC of a classifier G can be calculated as:

$$\hat{A} = \frac{S_0 - n_0(n_0+1)/2}{n_0 n_1} \quad (1)$$

where n_0 and n_1 are the numbers of positive and negative samples, respectively. Also, $S_0 = \sum r_i$, where r_i is the rank of the ith positive example in the ranked list where more positive examples are ranked higher.

We took a step forward to obtain more insight into the classification results and measured the performance of the classifier from a monetary aspect, i.e., we calculated the value created by adopting a classification algorithm in the decision-making process. In particular, we aggregated the total sales values in each of the four scenarios of classification (TP_m, TN_m, FP_m, FN_m) and defined monetary performance metrics with a similar formulation to the statistical metrics (see Table 3). For instance, the monetary precision is the fraction of the sales values correctly predicted as won.

Table 3. Statistical and monetary classifier performance metrics.

Statistical Metrics	
Notation	**Definition**
TP	True Positive: number of class 1 samples classified as 1
TN	True Negative: number of class 0 samples classified as 0
FP	False Positive: number of class 0 samples classified as 1
FN	False Negative: number of class 1 samples classified as 0
Metric	**Definition**
Precision	$TP/(TP+FP)$
Recall	$TP/(TP+FN)$
Accuracy	$(TP+TN)/(TP+TN+FP+FN)$
F1-Score	$2/(Recall^{-1}+Precision^{-1})$
Monetary Metrics	
Metric	**Definition**
$\text{Precision}_\text{m}$	$TP_m/(TP_m+FP_m)$
Recall_m	$TP_m/(TP_m+FN_m)$
Accuracy_m	$(TP_m+TN_m)/(TP_m+TN_m+FP_m+FN_m)$

2.3.2. XGBoost and LightGBM Classifiers

XGBoost, introduced by Chen and Guestrin [21], is a supervised classification algorithm that iteratively combines weak base learners into a stronger learner. With this algorithm, the objective function J is defined as

$$J(\Theta) = L(y, \hat{y}) + \Omega(\Theta) \tag{2}$$

where Θ denotes the model's hyperparameters. The training loss function L quantifies the difference between the prediction $\hat{y}$ and actual target value y. The regularization term Ω penalizes the complexity of the model with the L1 norm to smooth the learned model and avoid over-fitting. The model's prediction is an ensemble of k decision trees from a space of trees $\mathcal{F}$:

$$\hat{y}_i = \sum_{i=1}^{k} f_k(x_i), f_k \in \mathcal{F} \tag{3}$$

The objective function at iteration t for n instances can be simplified as:

$$J^{(t)} = \sum_{i=1}^{n} L(y_i, \hat{y}_i) + \sum_{k=1}^{t} \Omega(f_k) \tag{4}$$

where according to Equation (3), $\hat{y}_i$ can iteratively be written as

$$\hat{y}_i{}^{(t)} = \sum_{i=1}^{t} f_k(x_i) = \hat{y}_i{}^{(t-1)} + f_t(x_i) \tag{5}$$

The regularization term can be defined as

$$\Omega(f_k) = \gamma T + \frac{1}{2}\lambda \sum_{j=1}^{T} w_j^2 \tag{6}$$

where the coefficient γ is the complexity of each leaf. Also, T is the total number of leaves in the decision tree. To scale the weight penalization, λ can be tweaked. Using second-order Taylor expansion and assuming a mean square error (MSE) loss function, Equation (4) can be written as

$$J^{(t)} \approx \sum_{i=1}^{n} [g_i w_{q(x_i)} + \frac{1}{2}(h_i w_{q(x_i)})^2] + \frac{1}{2}\lambda \sum_{j=1}^{T} w_j^2 \tag{7}$$

Since each incident of data corresponds to only one leaf, according to [22], this can also be simplified as

$$J^{(t)} \approx \sum_{j=1}^{T} [(\sum_{i \in I_j} g_i) w_j + \frac{1}{2}(\sum_{i \in I_j} h_i + \lambda) w_j^2] + \gamma T \tag{8}$$

where I_j represents all instances of data in leaf j. As can be seen in Equation (8) minimizing the objective function can be transformed into finding the minimum of a quadratic function.

In an endeavor to reduce the computation time of the XGBoost algorithm, Ke, et al. proposed LightGBM [23]. The main difference between XGBoost and LightGBM is how they grow the decision tree (see Figure 3 for a high-level comparison). In XGBoost decision trees are grown horizontally (level-wise) while with LightGBM decision trees are grown vertically (leaf-wise). Importantly, this makes LightGBM an effective algorithm to handle datasets with high dimensionality.

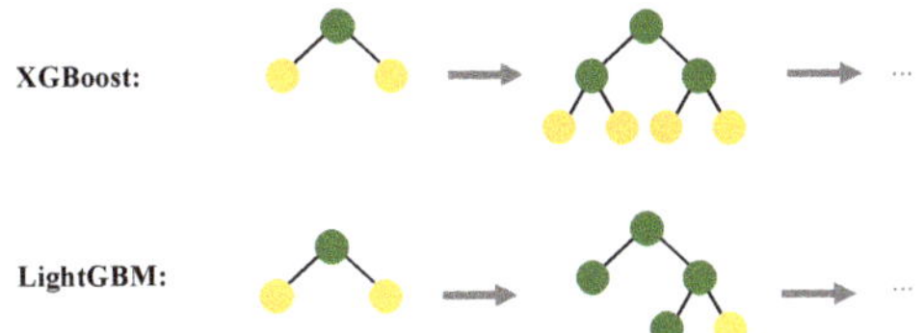

Figure 3. Comparison between XGBoost level-wise horizontal tree growth and LightGBM vertical leaf-wist tree growth.

2.3.3. Voting Ensemble

The main idea behind voting ensembles is to combine various classification models to balance out their individual classification errors. Voting ensembles predict the class label either by using most individual models' predictions (hard vote) [24] or averaging their predicted probabilities (soft vote) [25]. A voting ensemble was used to integrate the predictions of multiple iterations of both XGBoost and LightGBM classifiers with different parameterizations. Specifically, a soft-voting weighted average voting ensemble was used to combine the predictions for each model (Figure 4). A soft-voting ensemble is a meta-classifier model that computes its predicted probabilities y_i by takes the weighted average (w_j) probability predicted by each classifier (p_{ij}):

$$\hat{y}_i = \text{argmax}_i \sum_{j=1}^{m} w_j p_{ij} \tag{9}$$

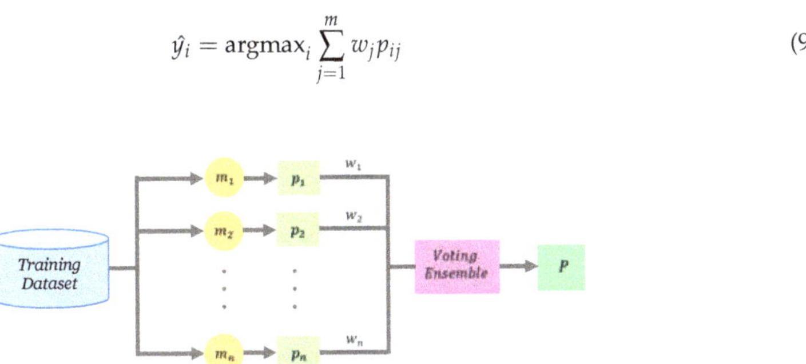

Figure 4. Voting Ensemble combines predictions p_i of multiple classifiers m_i using weighted average w_i to compute a final prediction P.

2.4. *Workflow and Pipelines*

Pipeline is defined as an executable workflow of data that is encapsulated in a series of steps. In this work, the proposed workflow consists of two main pipelines: (1) ML pipeline and (2) Prediction pipeline. All pipeline codes were custom-written in Python 3.7 on Microsoft Azure Machine-Learning Service [26] cloud platform. XGBoost v1.1 and LightGBM v2.3.1 libraries were integrated into Python to create ML classification models. The voting ensemble was created using the Microsoft Automated Machine-Learning tool [27].

2.4.1. Machine-Learning Pipeline

The main objective of the ML pipeline is to train predictive models on the closed sales opportunities data. As illustrated in Figure 5, there are four main steps in this pipeline:

(1) *Data Preparation*: Raw data of all closed sales opportunities are extracted from the CRM cloud database. Relevant features are selected for each sales record (see Table 1) and paired with their sales outcome (won/lost) as a class label. Please note that the user-entered probabilities are dropped to avoid biasing the model's predictions.

(2) *Feature Enhancement*: As described in Section 2.2, statistical analysis is performed on all categorical features to generate feature enhancement lookup tables for each of these categorical features (see Table 2). All lookup tables are stored back in the CRM cloud database. These tables are then appropriately merged back to the original selected features in the raw data.
(3) *Machine Learning*: A total number of 35 iterations of XGBoost and LightGBM classifiers with various parameterizations are trained on the data (Section 2.3.2). Eventually, all trained models are combined to generate a soft-voting ensemble classifier (Section 2.3.3).
(4) *Deploy Model to Cloud*: In the last step of the ML pipeline, the ensemble model is deployed as a web service using Azure ML. Azure ML platform supports creating a model's endpoint on Azure Kubernetes Service (AKS) cluster [28]. AKS enables request-response service with low latency and high scalability which makes it suitable for production-level deployments.

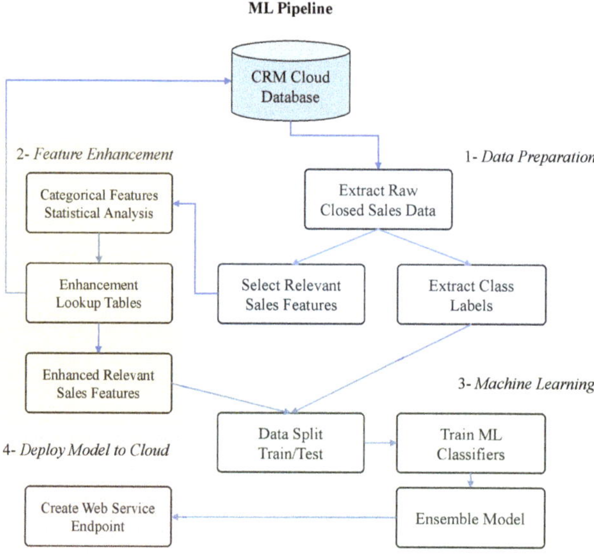

Figure 5. ML Pipeline: In four major steps the pipeline extracts and enhances sales data from a cloud database, trains an ensemble of ML classification models on the data, and eventually creates a cloud endpoint for the model.

2.4.2. Prediction Pipeline

The prediction pipeline, as illustrated in Figure 6, was designed to use the trained ML model and make predictions on the likelihood of winning new sales opportunities in four main steps:

(1) *Data Preparation*: All open sales records are extracted from the CRM cloud database. Relevant features are selected similar to the feature selection step in the ML pipeline. Please note that open sales opportunities are still active in the sales process and, hence, there is no final sales status (won/lost) assigned to them yet.
(2) *Feature Enhancement*: To make predictions on unseen data using the trained ML model, new input data needs to have a format similar to the data used to train the model. Therefore, all the previously stored lookup dictionaries are imported from the CRM cloud database and appropriately merged to the relevant features.
(3) *Machine-Learning Prediction*: The ensemble model created in the ML pipeline is called using its endpoint. The model makes predictions on the unseen sales data and assigns a probability of winning to each new opportunity.

(4) *Decision Boundaries*: All historical data on closed sales opportunities along with their ML predicted probabilities are grouped by the business segments (Healthcare, Energy, and Finance). Next, within each business segment, closed sales records are split into four quartiles based on their value. Then, the optimal decision boundary is calculated for each business segment's value quartile as described in Section 2.3.1. A total number of 12 decision boundaries are calculated (3 business segments × 4 quartiles). Eventually, all predicted probabilities and decision boundaries are stored back to the cloud database.

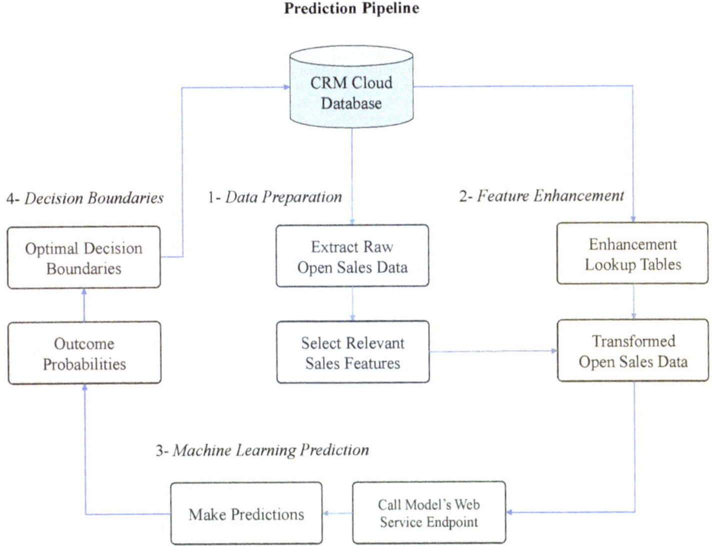

Figure 6. Prediction Pipeline: new sales opportunities data are transformed and enhanced, probability of winning is inferred using the trained ML model, and finally decision boundaries are optimized based on historical sales data.

3. Results

This section gives an overview of the proposed workflow's performance. The workflow was implemented in the CRM system of a global B2B consulting firm. The two pipelines were scheduled for automated runs on a recurring basis on the Azure ML platform. The ML pipeline was scheduled for a weekly rerun to retrain ML models on updated sales data and generate updated feature enhancement lookup tables. The prediction pipeline was scheduled for a daily rerun to calculate and store predictions for new sales opportunities.

3.1. Training the ML Model

A total number of 34 iterations of XGBoost and LightGBM were individually trained on the data and then combined in a voting ensemble classifier (see Section 2.3.3 for more details). The training accuracy was calculated using 10-fold cross-validation. The accuracy for each of the 35 iterations (with the last iteration being the voting ensemble classifier) is demonstrates in Figure 7a. Training accuracy for the top five model iterations are listed in Figure 7b. As expected, the voting ensemble had a slightly higher training accuracy compared to each individual classifier.

The voting ensemble classifier had a training accuracy of 81% (other performance metrics are listed in Table 4). On the train set, approximately 83% of the won sales and 79% of the lost sales were classified correctly (Figure 7d). For more insight into the training performance ROC curve (Figure 7c) is also illustrated. The area under the ROC curve (AUC) was equal to 0.87. In other words, this implies that a randomly drawn sample out of the train set has a 87% chance of being correctly classified by the trained model.

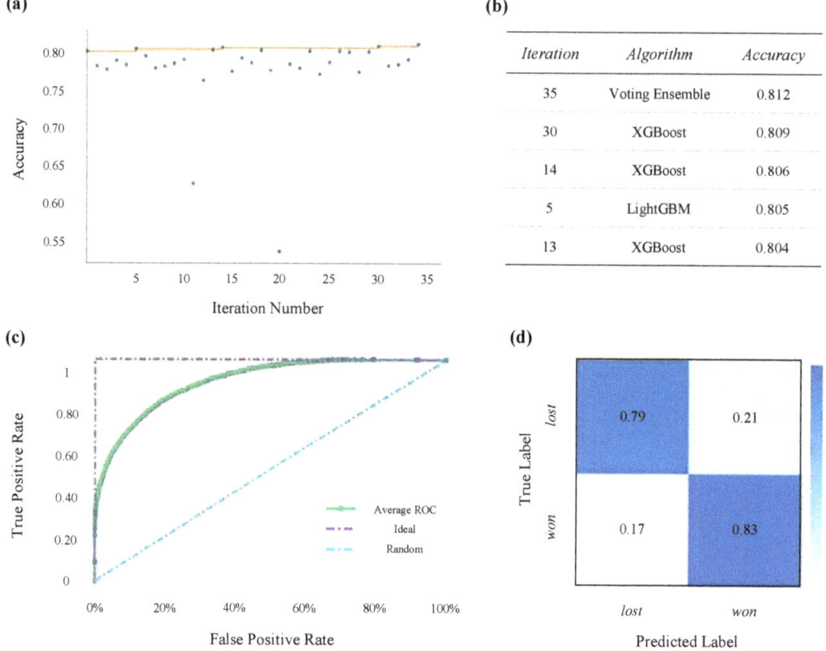

Figure 7. ML Training Results: (**a**) Training accuracy for all model iteration. (**b**) Accuracy of the top five iterated models sorted by the training accuracy. (**c**) ROC curve of the voting ensemble classifier. (**d**) Confusion matrix showing the four scenarios of classification for the voting ensemble model.

Table 4. Voting Ensemble Training Performance.

Metric	Value
Precision	0.81
Recall	0.83
Accuracy	0.82
AUC	0.87

3.2. Setting the Decision Boundaries

As explained in Section 2.4.2, statistical analysis of historical sales data is performed in each business segment (Healthcare, Energy, and Finance) to determine the decision boundaries. Specifically, the decision boundary was optimized for each of the four sales value quartiles of each business segment. The decision boundaries, demonstrated in Figure 8, ranged from 0.41 (Finance business segment—3rd value quartile) to 0.75 (Energy business segment—1st value quartile).

Interestingly, the decision boundaries were lower for sales opportunities with a higher monetary value which implies a more optimistic decision-making for more profitable opportunities. This sensible trend observed in the optimal decision boundaries provides more evidence to substantiate the idea of tailoring the boundaries uniquely to each business segment and value quartile due to their inherent decision-making differences.

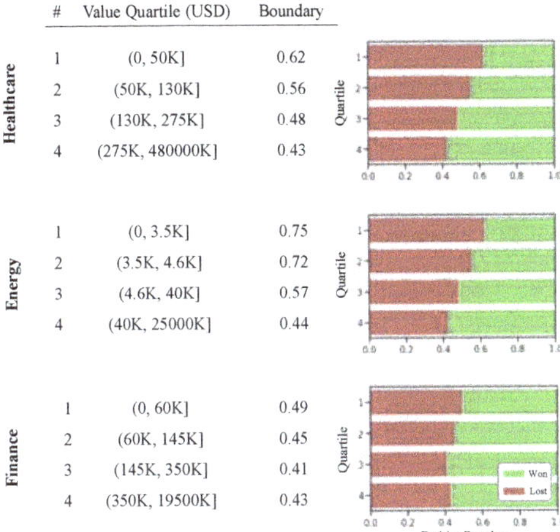

	#	Value Quartile (USD)	Boundary
Healthcare	1	(0, 50K]	0.62
	2	(50K, 130K]	0.56
	3	(130K, 275K]	0.48
	4	(275K, 480000K]	0.43
Energy	1	(0, 3.5K]	0.75
	2	(3.5K, 4.6K]	0.72
	3	(4.6K, 40K]	0.57
	4	(40K, 25000K]	0.44
Finance	1	(0, 60K]	0.49
	2	(60K, 145K]	0.45
	3	(145K, 350K]	0.41
	4	(350K, 19500K]	0.43

Figure 8. Decision Boundaries.

3.3. Model's Performance

The voting ensemble was used to make predictions on the unseen test set. In particular, after inferring the probability of winning for each sales opportunity, they were classified in accordance with a decision boundary corresponding to their business segment and value quartile. If the inferred probability of winning exceeded the decision boundary, a sales opportunity was classified to be won otherwise it was classified to be a lost opportunity. To make a concrete comparison between user-entered and ML predictions, statistical and monetary performance metrics were calculated for both approaches.

All four classification scenarios in the test set for both user-entered and ML prediction are depicted in Figure 9a. Qualitatively, the ML workflow made fewer false classifications (i.e., compare the true positive TP slice proportions in Figure 9a). More specifically, the ML workflow accurately classified 87% of the unseen sales data while the user-entered predictions only had an accuracy of 67%. In fact, all statistical performance metrics (precision, recall, and F1 score) were in favor of the ML predictions (see Table 5).

The performance of the user-entered and ML predictions was also compared with reference to the monetary metrics (see Section 2.3.1 for more details). As shown in Figure 9b, sales opportunities falsely predicted to be won by the ML workflow had considerably lower cumulative monetary value (compare the true positive FP_m slice proportions). This implies a lower monetary loss due to prediction error when using the ML predictions. Quantitatively, the monetary accuracy of the ML model was notably higher than the user-entered (90% versus 74%). Other monetary performance metrics are listed in Table 5.

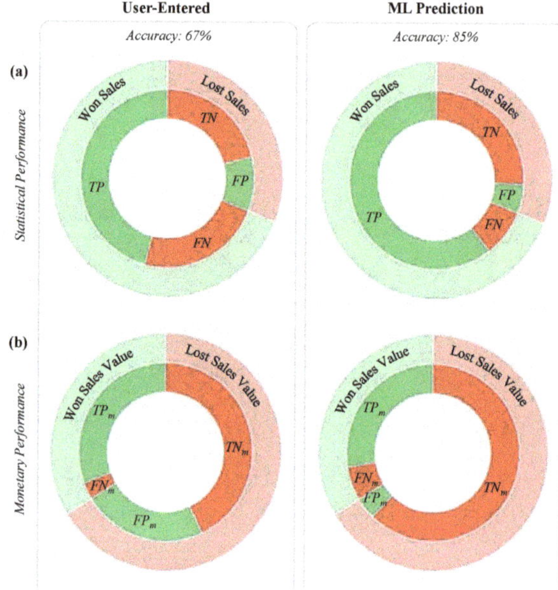

Figure 9. Test Set Results: (**a**) Statistical Performance: Actual outcome of all won (light green) and lost (light red) sales opportunities along with the corresponding predictions (solid green and red). (**b**) Monetary Performance: Cumulative contract value of won (light green) and lost (light red) sales opportunities along with the cumulative value of opportunities in each of the four classification scenarios (solid green and red). In both panels miss-matching colors indicates false classification.

Table 5. Test Set Performance Metrics.

Statistical Performance		
Metric	**User-Entered**	**ML**
Precision	0.82	0.92
Recall	0.66	0.87
F1-Score	0.73	0.89
Accuracy	0.67	0.85
Monetary Performance		
Metric	**User-Entered**	**ML**
Precision_m	0.57	0.87
Recall_m	0.91	0.82
Accuracy_m	0.74	0.90

3.4. Analysis of the Workflow Implementation

Similar to the previous section, a performance comparison between the user-entered and ML predictions was performed on a validation set. The validation set was collected while the workflow was implemented in the sales pipeline of a B2B consulting firm over a period of three months (see Section 2.1 for further details). A qualitative comparison in terms of statistical and monetary performance is presented in Figure 10. In the validation set, the ML workflow retained a substantially higher prediction accuracy (83% versus 63%). Also, there was an evident gap between the number of won sales misclassified by each approach (compare the true positive TP slices in Figure 10a).

The monetary accuracy of the ML predictions was marginally lower than the user-entered predictions (75% versus 77%). However, the cumulative value of the won sales opportunities correctly classified by the ML workflow was still considerably higher than the user-entered predictions (compare the true positive TP_m slices in Figure 10b). All performance metrics are listed in Table 6.

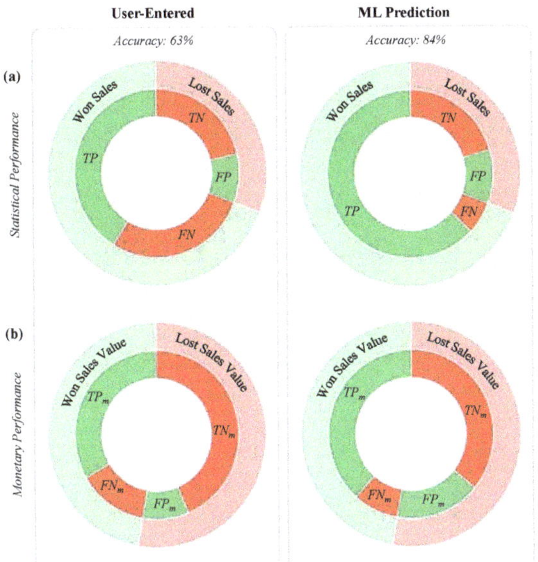

Figure 10. Validation Set Results: (**a**) Statistical and (**b**) Monetary performance of user-entered and ML predictions. Refer to Figure 9 caption for further explanations.

Table 6. Validation Set Performance Metrics.

Statistical Performance		
Metric	**User-Entered**	**ML**
Precision	0.82	0.85
Recall	0.59	0.92
F1-Score	0.70	0.87
Accuracy	0.63	0.83
Monetary Performance		
Metric	**User-Entered**	**ML**
Precision_m	0.79	0.71
Recall_m	0.72	0.82
Accuracy_m	0.77	0.75

4. Conclusions

In this paper, we proposed a novel ML workflow implemented on a cloud platform for predicting the likelihood of winning sales opportunities. With this approach, sales data were extracted from the CRM cloud database and then improved by an extensive feature enhancement approach. The data was then used to train an ensemble of probabilistic classification models in parallel. The resulting meta classification model was then used to infer the likelihood of winning new sales opportunities. Lastly, to maximize the interpretability of the predictions, optimal decision boundaries were calculated by performing statistical analysis on the historical sales data.

To inspect the effectiveness of the ML approach, it was deployed to a multi-business B2B consulting firm for over three months. The performance of the ML workflow was compared with the user-entered predictions made by salespersons. Standard statistical performance metrics confirmed that by far the ML workflow provided superior predictions. From a monetary standpoint, the value created from decision-making was also higher when incorporating the ML workflow.

The proposed ML workflow is a cloud-based solution that can readily be integrated into the existing cloud-based CRM system of enterprises. On top of that, this workflow is highly sustainable and scalable since it relies on cloud computing power instead of on-premise computing resources. Although our proposed workflow is mainly built around Azure ML platform, future work can explore implementing this workflow on other cloud computing resources such as Amazon web services, Google cloud platform, etc.

A potential issue with the proposed workflow is handling the scenario of imbalanced dataset. An imbalanced dataset is characterized by having more instances of a certain class compared to others [29]. In our problem, this would translate to a dataset that has more lost sales record than won (or vice versa). For instance, consider Energy or Finance business segments (Figure 1) in our data set where the number of won and lost sales records are unbalanced. Solutions to deal with an imbalance problem at the data-level involves oversampling the smaller class or undersampling the larger class [30–32]. Galar et al. [33] showed that combining random undersampling techniques with ensemble models stands out among other data-level solutions. In future work, we hope to explore this idea by incorporating an undersampling technique to the existing ensemble models in the workflow.

The results obtained in this work suggest a data-driven ML solution for predicting the outcome of sales opportunities is a more concrete and accurate approach compared to salespersons' subjective predictions. However, it is worth mentioning that ML solutions should not be overwhelmingly used to rule out sensible or justifiable sentiments of salespersons in evaluating a sales opportunity. A data-driven approach, such as the workflow presented in this work, can provide a reliable reference point for further human assessments of the feasibility of a sales opportunity.

Funding: This research received no external funding.

Acknowledgments: The author would like to thank Saeed Fotovat for his valuable suggestions and comments.

Conflicts of Interest: The authors declare no conflict of interest.

References

1. Monat, J.P. Industrial sales lead conversion modeling. *Mark. Intell. Plan.* 2011. Available online: https://www.emerald.com/insight/content/doi/10.1108/02634501111117610/full/html (accessed on 5 August 2020). [CrossRef]
2. Bohanec, M.; Borštnar, M.K.; Robnik-Šikonja, M. Integration of machine learning insights into organizational learning: A case of B2B sales forecasting. In *Blurring the Boundaries through Digital Innovation*; Springer: Berlin/Heidelberg, Germany, 2016; pp. 71–85.
3. Matthies, B.; Coners, A. Double-loop learning in project environments: An implementation approach. *Expert Syst. Appl.* **2018**, *96*, 330–346. [CrossRef]
4. Duran, R.E. Probabilistic sales forecasting for small and medium-size business operations. In *Soft Computing Applications in Business*; Springer: Berlin/Heidelberg, Germany, 2008; pp. 129–146.
5. Bohanec, M.; Robnik-Šikonja, M.; Borštnar, M.K. Organizational learning supported by machine learning models coupled with general explanation methods: A Case of B2B sales forecasting. *Organizacija* **2017**, *50*, 217–233. [CrossRef]
6. Ingram, T.N.; LaForge, R.W.; Schwepker, C.H.; Williams, M.R. *Sales Management: Analysis and Decision Making*; Routledge: Los Angeles, CA, USA, 2015.
7. Davis, D.F.; Mentzer, J.T. Organizational factors in sales forecasting management. *Int. J. Forecast.* **2007**, *23*, 475–495. [CrossRef]
8. Armstrong, J.S.; Green, K.C.; Graefe, A. Golden rule of forecasting: Be conservative. *J. Bus. Res.* **2015**, *68*, 1717–1731. [CrossRef]

9. Xu, X.; Tang, L.; Rangan, V. Hitting your number or not? A robust & intelligent sales forecast system. In Proceedings of the 2017 IEEE International Conference on Big Data (Big Data), Boston, MA, USA, 11–14 December 2017; pp. 3613–3622.
10. Davis, J.; Fusfeld, A.; Scriven, E.; Tritle, G. Determining a project's probability of success. *Res. Technol. Manag.* **2001**, *44*, 51–57. [CrossRef]
11. De Oliveira, M.G.; Rozenfeld, H.; Phaal, R.; Probert, D. Decision making at the front end of innovation: The hidden influence of knowledge and decision criteria. *R D Manag.* **2015**, *45*, 161–180. [CrossRef]
12. Yan, J.; Gong, M.; Sun, C.; Huang, J.; Chu, S.M. Sales pipeline win propensity prediction: A regression approach. In Proceedings of the 2015 IFIP/IEEE International Symposium on Integrated Network Management (IM), Ottawa, ON, Canada, 11–15 May 2015; pp. 854–857.
13. Lambert, M. Sales Forecasting: Machine Learning Solution to B2B Sales Opportunity Win-Propensity Computation. Ph.D. Thesis, National College of Ireland, Dublin, Ireland, 2018.
14. Bohanec, M.; Kljajić Borštnar, M.; Robnik-Šikonja, M. Feature subset selection for B2B sales forecasting. In Proceedings of the 13th International Symposium on Operational Research, SOR, Bled, Slovenia, 23–25 September 2015; Volume 15, pp. 285–290.
15. Abdi, H. Coefficient of variation. *Encycl. Res. Des.* **2010**, *1*, 169–171.
16. De Maesschalck, R.; Jouan-Rimbaud, D.; Massart, D.L. The mahalanobis distance. *Chemom. Intell. Lab. Syst.* **2000**, *50*, 1–18. [CrossRef]
17. Abbott, D. Foreword 2 for 1st Edition. In *Handbook of Statistical Analysis and Data Mining Applications*; Academic Press: Cambridge, MA, USA, 2018; pp. xv–xvi.
18. Bishop, C.M. *Pattern Recognition and Machine Learning*; Springer: Berlin/Heidelberg, Germany, 2006.
19. Bewick, V.; Cheek, L.; Ball, J. Statistics review 13: Receiver operating characteristic curves. *Crit. Care* **2004**, *8*, 1–5.
20. Hand, D.J.; Till, R.J. A simple generalisation of the area under the ROC curve for multiple class classification problems. *Mach. Learn.* **2001**, *45*, 171–186.
21. Chen, T.; Guestrin, C. Xgboost: A scalable tree boosting system. In Proceedings of the 22nd ACM Sigkdd International Conference on Knowledge Discovery and Data Mining, San Francisco, CA, USA, 13–17 August 2016; pp. 785–794.
22. Zhang, D.; Qian, L.; Mao, B.; Huang, C.; Huang, B.; Si, Y. A data-driven design for fault detection of wind turbines using random forests and XGboost. *IEEE Access* **2018**, *6*, 21020–21031. [CrossRef]
23. Ke, G.; Meng, Q.; Finley, T.; Wang, T.; Chen, W.; Ma, W.; Ye, Q.; Liu, T.Y. Lightgbm: A highly efficient gradient boosting decision tree. In Proceedings of the Advances in Neural Information Processing Systems, Long Beach, CA, USA, 4–9 December 2017; pp. 3146–3154.
24. Ruta, D.; Gabrys, B. Classifier selection for majority voting. *Inf. Fusion* **2005**, *6*, 63–81. [CrossRef]
25. Kuncheva, L.I.; Rodríguez, J.J. A weighted voting framework for classifiers ensembles. *Knowl. Inf. Syst.* **2014**, *38*, 259–275. [CrossRef]
26. Barga, R.; Fontama, V.; Tok, W.H. Introducing microsoft azure machine learning. In *Predictive Analytics with Microsoft Azure Machine Learning*; Springer: Berlin/Heidelberg, Germany, 2015; pp. 21–43.
27. Feurer, M.; Klein, A.; Eggensperger, K.; Springenberg, J.; Blum, M.; Hutter, F. Efficient and robust automated machine learning. In Proceedings of the Advances in Neural Information Processing Systems, Montreal, QC, Canada, 7–12 December 2015; pp. 2962–2970.
28. Barnes, J. Azure machine learning. In *Microsoft Azure Essentials*, 1st ed.; Microsoft Press: Redmond, WA, USA, 2015.
29. Sun, Y.; Wong, A.K.; Kamel, M.S. Classification of imbalanced data: A review. *Int. J. Pattern Recognit. Artif. Intell.* **2009**, *23*, 687–719. [CrossRef]
30. Chawla, N.V.; Bowyer, K.W.; Hall, L.O.; Kegelmeyer, W.P. SMOTE: Synthetic minority over-sampling technique. *J. Artif. Intell. Res.* **2002**, *16*, 321–357.
31. Chawla, N.V.; Japkowicz, N.; Kotcz, A. Special issue on learning from imbalanced data sets. *ACM SIGKDD Explor. Newsl.* **2004**, *6*, 1–6. [CrossRef]

32. Zhou, Z.H.; Liu, X.Y. Training cost-sensitive neural networks with methods addressing the class imbalance problem. *IEEE Trans. Knowl. Data Eng.* **2005**, *18*, 63–77. [CrossRef]
33. Galar, M.; Fernandez, A.; Barrenechea, E.; Bustince, H.; Herrera, F. A review on ensembles for the class imbalance problem: Bagging-, boosting-, and hybrid-based approaches. *IEEE Trans. Syst. Man Cybern. C Appl. Rev.* **2011**, *42*, 463–484. [CrossRef]

Article

Bus Travel Time: Experimental Evidence and Forecasting

Antonio Comi * and Antonio Polimeni

Department of Enterprise Engineering, University of Rome Tor Vergata, 00118 Rome, Italy; antonio.polimeni@uniroma2.it

* Correspondence: comi@ing.uniroma2.it; Tel.: +39-06-7259-7061

Received: 28 July 2020; Accepted: 26 August 2020; Published: 28 August 2020

Abstract: Bus travel time analysis plays a key role in transit operation planning, and methods are needed for investigating its variability and for forecasting need. Nowadays, telematics is opening up new opportunities, given that large datasets can be gathered through automated monitoring, and this topic can be studied in more depth with new experimental evidence. The paper proposes a time-series-based approach for travel time forecasting, and data from automated vehicle monitoring (AVM) of bus lines sharing the road lanes with other traffic in Rome (Italy) and Lviv (Ukraine) are used. The results show the goodness of such an approach for the analysis and reliable forecasts of bus travel times. The similarities and dissimilarities in terms of travel time patterns and city structure were also pointed out, showing the need to take them into account when developing forecasting methods.

Keywords: travel time forecasting; time series; bus service; transit systems; sustainable urban mobility plan; bus travel time

---

1. Introduction

Travel time plays a key role in assuring the reliability and quality of service in a transit system. The use of this variable ranges from operation planning (e.g., for short- and long-term planning) to service monitoring (e.g., for real-time information). Moreover, in large historical cities (e.g., Rome, Italy), the influence of traffic congestion on travel time variability has been pointed out. It is linked, on the one hand, to the city structure and, on the other hand, with the growing number of private and freight vehicles travelling into the city [1–3]. Therefore, transit operators have to take into consideration the travel time variability when they design new lines or when they work to improve existing lines. Usually, in cities, transit services share the road lanes with other vehicles; subsequently, their travel times are highly impacted by congestion, and their temporal patterns could be similar to private/freight ones, as emerged from some surveys carried out around the world [4,5], where seasonality and trend/cycle components were revealed. Time series analysis captures this pattern [6].

Therefore, the availability of reliable travel time forecasting is a relevant attribute for transit operators [6–10] to use when designing or updating service timetables. This can also contribute to attracting more passengers and increasing their satisfaction [11,12]. Moreover, in order to provide accurate information to the passenger, models and procedures must be developed to forecast travel time.

According to the Guidelines for Developing and Implementing a Sustainable Urban Mobility Plan (SUMP; [13]), SUMP has to improve urban accessibility as well as to offer users high-quality and sustainable mobility services from/to the study area. It regards the needs of the *functioning city* and its hinterland rather than a municipal administrative region. Furthermore, SUMP has to foster a balanced development of all relevant transport modes, while encouraging a shift towards more sustainable modes (e.g., transit). The plan puts forward an integrated set of technical, infrastructure, policy-based, and soft measures to improve performance and cost-effectiveness with regard to the declared goal and specific objectives [13]. Among the topics that are typically addressed, one of the most relevant is public transport, for which it is requested to present a strategy to enhance the quality, security, integration and accessibility of public transport services, covering infrastructure, rolling stock, and services. Public transport is a good way to reduce congestion and environment- and health-harming emissions in urban areas, especially when they run on alternative, cleaner fuels. Therefore, information on the state of the public network and reliable timetables and service schedules may be an effective tool for improving the quality and effectiveness of services perceived by users ([14]) and, hence, for diverting people to public transport modes.

Currently, one of the new challenges facing urban planners is to find solutions that can reduce the impacts of urban mobility without penalising passengers' mobility needs. Therefore, since suitable time estimates are needed for designing timetables and schedule services on the different routes, the development of travel time forecasting methodologies (including models) has to point out the complexity of urban systems as well as city sizes and characteristics. In fact, cities can differ from each other in both mobility/transport patterns and traffic conditions, as well as in other factors including geographical, environmental, demographic and socioeconomic conditions, cultural backgrounds, and institutional and legal frameworks. Hence, an incoming challenge is to develop performing forecasting methods and test their transferability [14,15]. Therefore, an outline of city similarities or dissimilarities with respect to the travel time of transit services could be useful for supporting the development and implementation of project/scenario actions. Such analysis could be a preguideline for planners in an ex-ante assessment. It can verify whether the experiment results and forecasting methods used in a city match those obtained from other cities in the way of defined goals (e.g., shifting users to more sustainable transit services) and identify factors that need in-depth and specific investigations.

Using some comparable surveys carried out in two cities that are very different in terms of spatial and economic patterns, the paper highlights the similarity and dissimilarity that exist in bus travel time and points out an easy-to-apply methodology for forecasting. Such a tool can help transit agencies achieve further advances in transit systems, both from the travellers' (i.e., transit trip planners) and operators' perspective (i.e., decision support system for operations control). The paper thus aims at identifying if the results obtainable by travel time forecasting methods can be easily transferred from one city to another.

Information about travel time, thus, benefits operators and passengers [16]. It assists operators in defining optimal slack times to maximise the on-time arrival performance of buses [17], in determining the reliability of systems [18], and in defining timetables [19]. Accuracy in travel time forecasting for defining service timetables and schedules can contribute to increasing the perceptions of service quality and help users in their decision-making about departure time and route choice [20–22]. This is found to be as valuable [23], or even more valuable, than a reduction in travel time [24].

In this paper, a time-series approach is applied to investigate bus travel times; the analysed data refer to the automatic vehicle location (AVL) of some transit lines in the cities of Rome (Italy, [5]) and Lviv (Ukraine, [25]). The first objective is to correlate bus travel time with general traffic patterns and other explanatory variables. Once the bus travel time has been analysed, the second objective is to provide a travel time forecasting procedure based on time series and to point out if similarities exist among travel time patterns in cities that are quite different.

This paper is organised as follows. Section 2 briefly reviews the current literature on bus travel time forecasting methods. Section 3 summarises the available data, while Section 4 describes the methodology used and synthetises the data available, the analyses performed, and the results obtained. Finally, Section 5 draws conclusions and the road ahead.

2. Literature Review

As said, travel time and, of course, onboard loads [26,27] are two of the most commonly used performance indicators in public transport systems. In particular, travel time (or commercial speed) is generally proposed as one of the fundamental parameters for assessing the effectiveness of the transport service. On the other hand, providing users with accurate and reliable travel forecasts can be a valid driver in attracting new demand and, therefore, favouring modal shifting.

In the following, unless otherwise specified, travel time between two successive time points is defined as the total time taken between departure (or passage) from one time-point to restart (or passage) from the next time-point. The definition of travel time is shown in Figure 1.

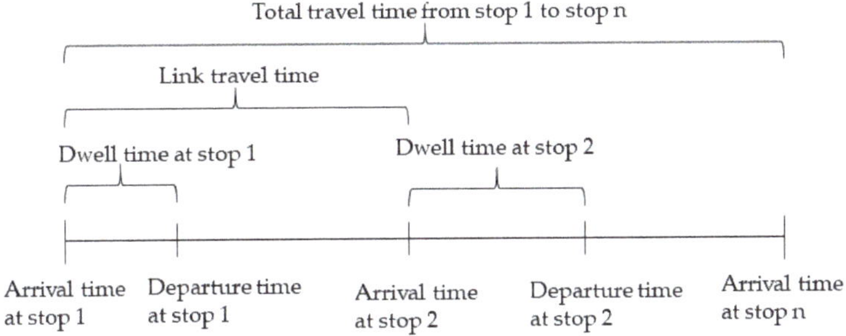

Figure 1. Notation on bus travel time [28].

Travel time forecasts can be classified as short term/real-time forecasts and long-term forecasts. The difference between the two types is mainly due to their forecast horizons. The short-term travel time forecast aims to predict the travel time on the forecast horizon as equal to or less than a time value T [22], in which T could be defined case by case, e.g., 15 min. The long-term travel time forecast aims to predict the travel time as the longest forecast horizon of T, which could be the following day, week or year [29,30]. Long-term forecasting uses only historical average traffic condition data to predict future traffic status and travel time. Figure 2 illustrates the definition of travel time estimation and forecast/prediction.

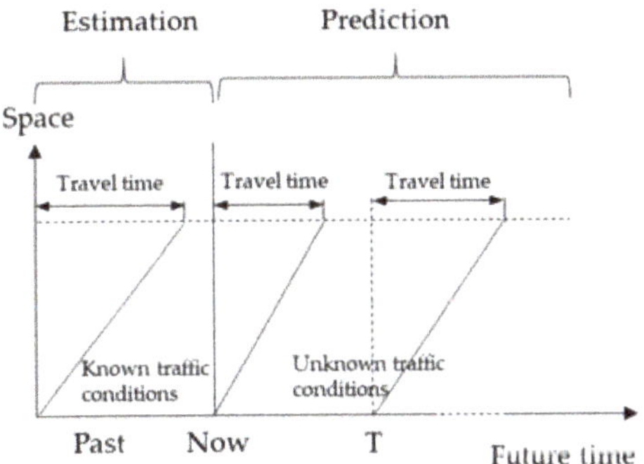

Figure 2. Estimation and prediction notations on bus travel time [28].

Therefore, given the values of the travel time observed up to time T, the forecast/prediction at time T of the future realisation can be of two types:

- historical: the forecasting model obtained is applied to a future situation, for example, the following month or the following year, in the context of operational planning.
- real-time: current data, possibly combined with historical, are used to predict future values, for example, in the context of operations control.

Real-time forecasting can, in turn, follow the one-step-ahead forecasting approach, i.e., at time $T + 1$, realisations up to time $T + 1$ are used to forecast the value of the variable at time $T + 2$. In the event that the weights to be attributed to the previous creations decrease over time, there is the approach known in the literature as "exponential smoothing" [31].

In transit systems, travel time (*TT*) forecasts are used both for operations and monitoring planning [32–37], and different methods and models have been developed both for long- and short-term. Below, the literature is reviewed, and the pros and cons of each proposed approach are pointed out. Long-term travel times that are forecast for bus services [38–40] have been studied through *regression methods* and *time series methods*. Regression methods explain travel time (the dependent variable) through a set of independent variables (e.g., loads, street characteristics). To point out that nonlinear relationships between independent and dependent variables can exist [38], more complex models were developed, e.g., k-nearest neighbour regression, support vector regression, project pursuit regression, and artificial neural networks. Although such methods do not give satisfactory results when traffic conditions are not stable, they are largely applied [38–40] because they reveal which independent variables are relevant or most important for the reproduction/forecast of travel times.

Time-series-based methods focus on the relationship between the variables to be predicted (travel time) through the analysis of historical data [41–44]. Their strength consists of the high calculation speed due to the simple formulation of the analysis algorithm and the small number of operating variables of the service: only the travel times of the bus relative to the instant in time to which they were found. They allow the structure of travel time variability to be highlighted and the effects over time (e.g., hours of the day, day of the week, time of year), which are relevant in the links where the buses travel alongside other traffic components [5,45], to be revealed.

Deepening studies developed for short-term forecasting, these works can be classified into time-series [46,47], regression models [8,48], artificial neural networks (ANNs) [8,49–52], Kalman filter [52–55], and nonparametric regression models (NPRs) [47,56–58]. Time series models, such as autoregressive integrated moving average (ARIMA) models and exponential smoothing models, can forecast based on historical values. Large variations in historical data could lead to significant differences between observations and forecasts because time series models depend on the transferability of historical trends (patterns) to forecast trends [50]. Regression models are used to predict travel time through linkage with context-specific independent variables (e.g., traffic conditions, road link characteristics). Given that the explanatory variables must be statistically independent of each other and many of the variables relating to transport systems are highly correlated [52], ANN models have been developed. In fact, ANN models allow complex nonlinear relationships to be pointed out and can also generate better results than other models such as Kalman filter-based methods, smoothing models, historical profiles, and real-time profiles [51]. However, ANN models take longer computational time for the training process than other models.

3. Forecasting Methodology and Data Analysis

The collected data are related to two lines in the city of Rome (Italy; Lines A and B) and one line (Line C) in the city of Lviv (Ukraine). Table 1 reports the characteristics of the investigated bus service lines (Rome and Lviv). Figure 3 shows the observed travel times during working days (i.e., from Monday to Friday). The travel time for each bus line is analysed in two cases: *to the city centre* and *from the city centre*. According to the time of the day, we can observe a similarity in the shape of the data if travel is to or from the city centre: two peak hours present in the mornings and afternoon. Particularly, for the investigated bus lines, the variance is higher in the afternoon peak hour than in the morning (e.g., for Line A, direction to city centre: 229,800 sec^2 vs. 160,284 sec^2; direction from city centre: 155,034 sec^2 vs. 110,413 sec^2). Moreover, given that the roads are congested during peak hours and less congested during nonpeak hours, and, subsequently, travel time is longer in rush hours than in nonpeak hours, time series allow these patterns to be captured. In the following sections, time-series components of bus travel time data are studied.

Table 1. The investigated bus lines.

Line	City	ATT (*) [minutes]	ATT (#) [minutes]	TL [km]	Peak-Hour Headway [minutes]	Off-Peak Hour Headway [minutes]	N	T [weeks]
A	Rome	39.5	41.7	11	20	30	37	8
B	Rome	57.5	58.4	23	15	30	60	12
C	Lviv	39.2	41.5	19	-	-	17	10

ATT: daily average travel time; TL: travel length; N: one-way number of stops; T: observation period. (*) to the city centre; (#) from the city centre.

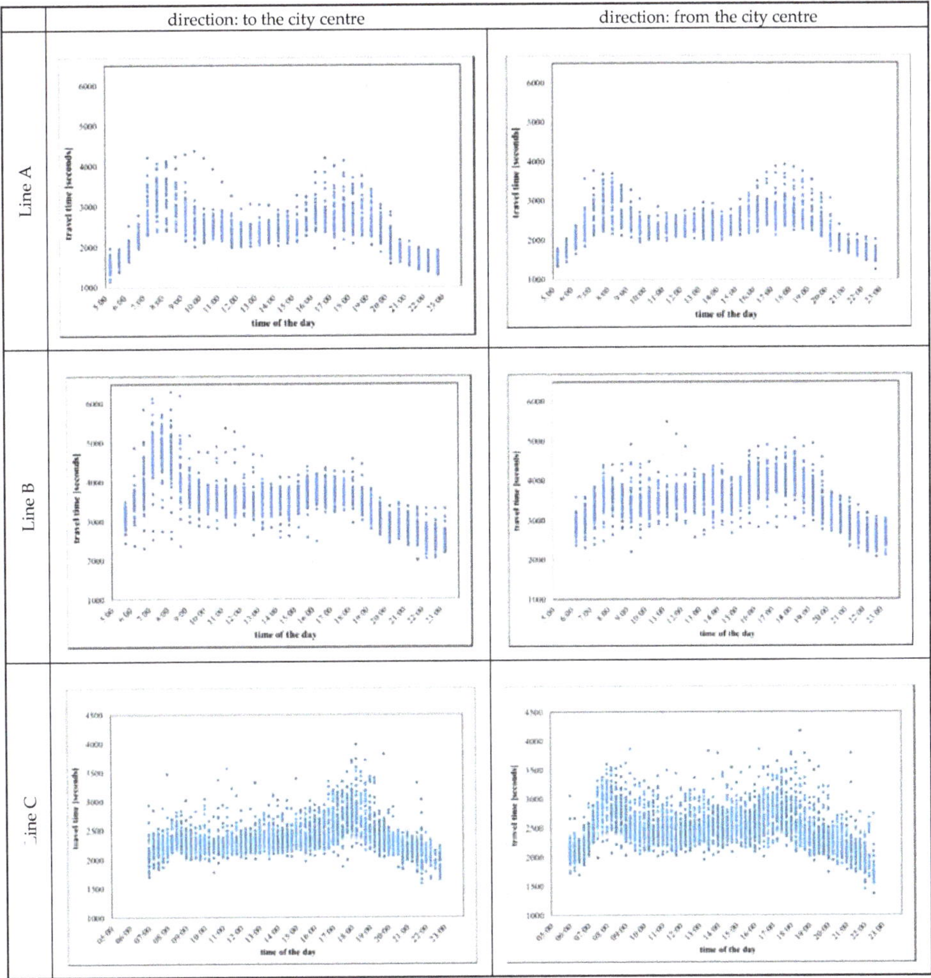

Figure 3. Analysed bus lines: hourly fluctuation of travel time during working days.

4. Bus Travel Time Analysis and Forecasting

4.1. Bus Travel Time Analysis

Given a bus line, the bus travel time (*TT*) from a terminal to another is assumed to be the sum of the running time (*RT*) between successive stops and the dwelling time at stops (*DW*):

$$TT = \sum_i RT_i + \sum_i DW_i \tag{1}$$

with RT_i the running time between stop i and successive one i+1, and DW_i the dwelling time spent at stop i.

Running time and dwelling time depends on a set of determinants. Running time is a function of speeds (related to flow composition and link flow), link characteristics (infrastructural and functional), context conditions (e.g., weather conditions; [59]), while dwelling time depends on on-board flow, alighting and boarding users, and bus features (e.g., number of doors, lift operations). Therefore, the analysis of bus travel time variability has to be investigated by capturing the fluctuations of these determinants. This pattern can be pointed out by time-series analysis.

A time series Y_t can be composed of three components: a trend-cycle component T_t, a seasonal component S_t, and a remainder component E_t that contains anything else in the time series. Therefore, assuming an additive relation among the three above components, the time series can be expressed as

$$Y_t = f(T_t, S_t, E_t) = T_t + S_t + E_t \tag{2}$$

Once the output of time series decomposition is used for forecasting, forecast accuracy has to be evaluated. An approach to evaluating the accuracy of forecasts relies on the application of the model on new data that were not used in the model calibration [31]. To do this, the available data are divided into two sets, *training* and *test* data. The training data is used to set up the forecasting model (i.e., identification of time-series components) and the test data is used to evaluate its accuracy (e.g., one week ahead). Since the test data is not used in determining the model, it should provide a reliable indication of how well the model is able to forecast by using new data. With this approach, the forecast error (e_i) is assumed to be the difference between an observed value and its forecast, as follows:

$$e_i = Y_i - \hat{Y}_i \tag{3}$$

where Y_i denotes the ith observation and $\hat{Y}_i$ denotes a modelled value of Y_i.

The most known accuracy measures [31] are MAE (mean absolute error), RMSE (root mean squared error), and MAPE (mean absolute percentage error). MAE and RMSE are scale-dependent measures and are based on *absolute errors* $|e_i|$ or *squared errors* $(e_i)^2$:

$$\begin{aligned} \text{mean absolute error}: \ \text{MAE} &= \text{mean}(|e_i|) \\ \text{root mean squared error}: \ \text{RMSE} &= \sqrt{\text{mean}\left(e_i^2\right)} \end{aligned} \tag{4}$$

On the other hand, MAPE has the advantage of being *scale-independent*:

$$\text{mean absolute percentage error}: \ \text{MAPE} = \text{mean}\left(|p_i|\right) \tag{5}$$

with $p_i = 100 \cdot e_i / Y_i$, which measures the percentage error.

4.2. Bus Travel Time Forecasting

As mentioned above, the used data refer to observed values of three bus lines during some working days of 10 consecutive weeks in Rome and Lviv. The last two weeks were set up as the test set, while the remaining weeks as the training set. Therefore, the time-series period was set in a week (five working days).

The time series decomposition (Figures 4 and 5) was performed through seasonal and trend decomposition using the loess (STL; [60,61]) method implemented in R software [31]. The results analysis can be pointed out, and it shows similarities with patterns revealed in other urban contexts ([32,45,62]):

- trends/cycles (T): small differences between maximum and minimum values, i.e., less than 5% for Line A, and about 2% for Lines B and C;
- weekly seasonality (S; Figure 6): the effects emerge for all days, with a periodic shape; differences emerge among the first and last days of the week (i.e., Monday/Tuesday vs. Thursday/Friday);
- daily seasonality (S; Figure 6):
    - ○ quite relevant for different hours of the day; as expected, it is influenced by the variance of traffic flows (i.e., buses share the lanes with other traffic components and bus travel time is influenced);
    - ○ quite different for routes to or from the city centre; higher values were revealed in the morning due to high concentrations of constrained trip arrivals (e.g., systemic trips, such as to work or school); on the other hand, the effects are more spread along the hours in the afternoon.
- remainder (E): low contributions in terms of variance of about 29% (Line A) and 21% (Line B) for Rome, while it is quite high in Lviv (Line C); it reflects the singular variability revealed in some days because of chance events concentrated in time and space rather than structural factors (Table 2).

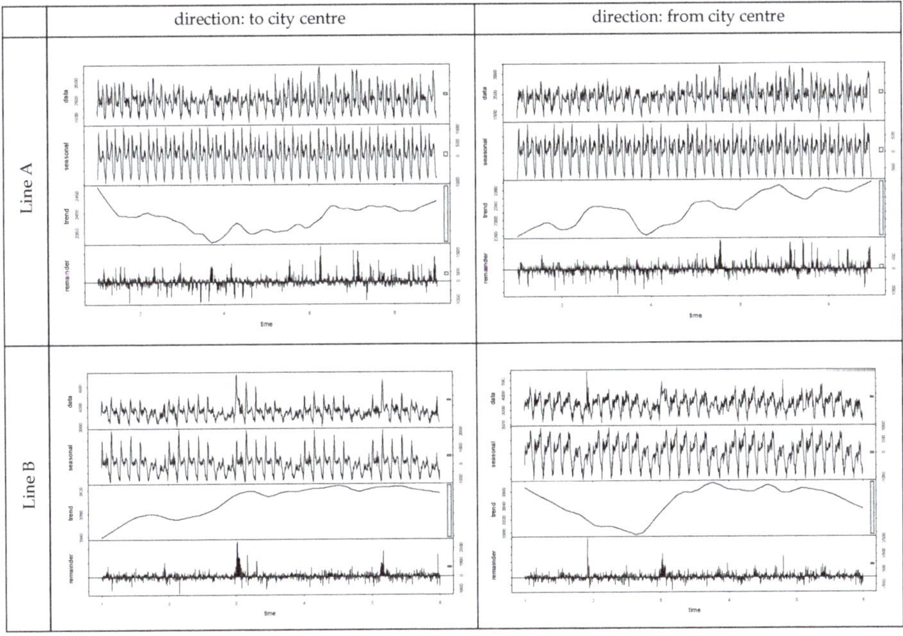

Figure 4. Travel time STL decomposition with weekly period (8 weeks): Rome.

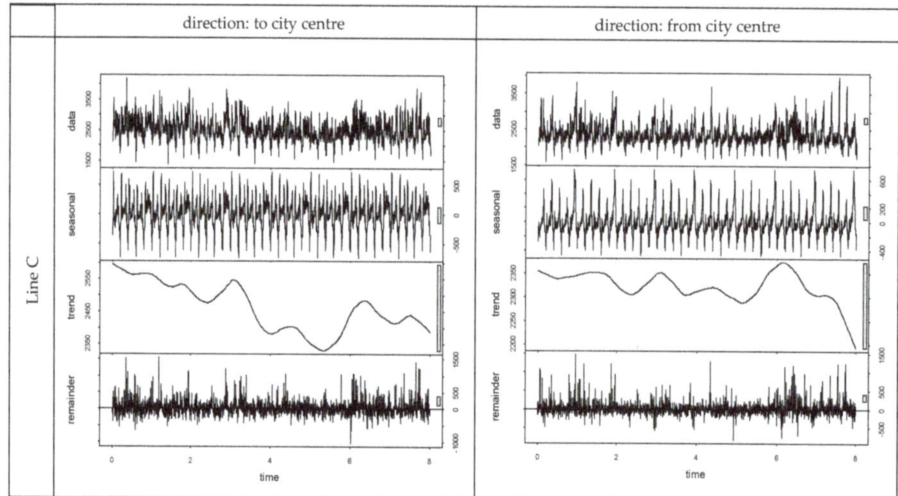

Figure 5. Travel time STL decomposition with weekly period (8 weeks): Lviv.

The capability to reproduce the observed data is then tested by using trend/cycle (T_t) and daily/hourly seasonality (S_t; i.e., time-series systematic components). The modelled error e can be assumed to be the remainder E (i.e., $e \equiv E$).

Figure 7 shows an example of a comparison between observed and modelled travel time for the test set, while Table 3 reports the accuracy for the investigated period. As synthetised by MAPE (smaller than 8%), the systemic component of travel time (i.e., trend/cycle and seasonality) allows the main part of variance to be explained. It means an average error in reproducing travel time less than 3 min (and about 1 min for Line B from the city centre) for the investigated lines. These accuracy metrics have the same magnitude as those we can find in the literature ([19,27,45,54]).

The results of these analyses refer to the improvement of long-term travel time forecasting, but they can suggest new research opportunities for short-term forecasting, as also suggested by Cristobal et al. [8], who proposed short-term travel bus methods based on the similarity between historical ones.

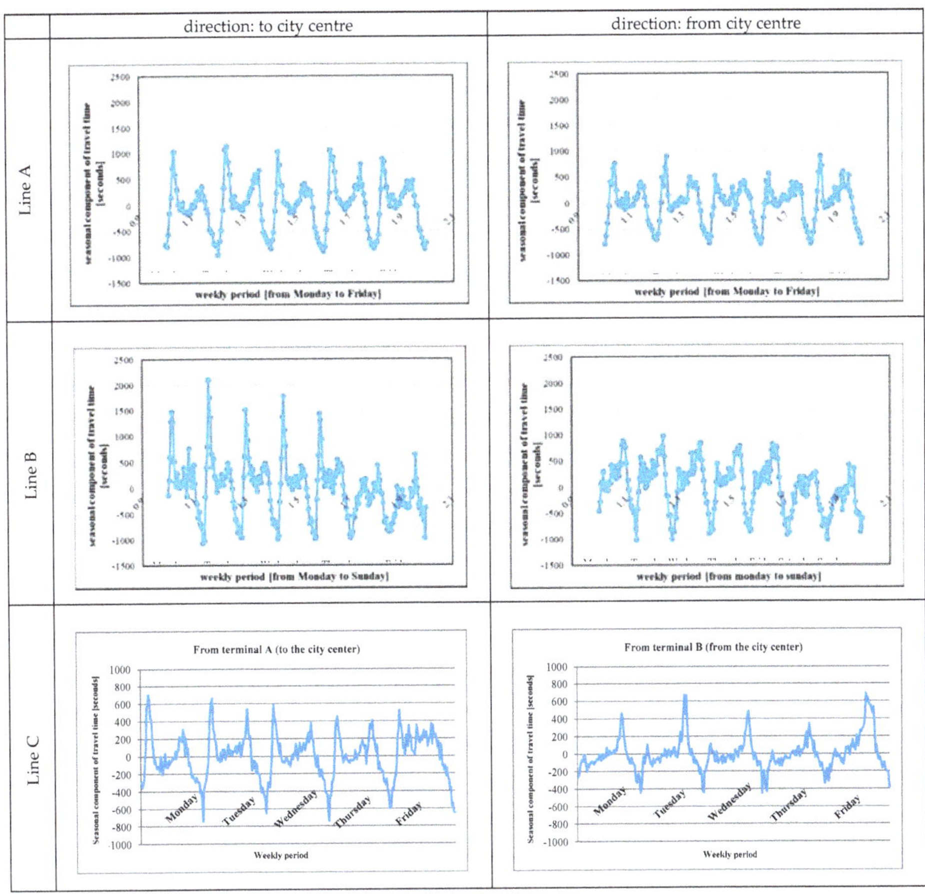

Figure 6. Seasonal components of the analysed bus lines.

Table 2. Variance (σ^2) and mean (μ) of the observed travel time (Y) and the remainder (E).

	Direction: To the City Centre	Direction: From the City Centre
Line A	σ^2 [Y] = 297,754	σ^2 [Y]= 198,696
	μ [Y] = 2412 s	μ [Y] = 2357 s
	σ^2 [E] = 83,436	σ^2 [E] = 58,329
	μ [E] = 30 s	μ [E] = 23 s
Line B	σ^2 [Y] = 405,878	σ^2 [Y] = 248,132
	μ [Y] = 3435 s	μ [Y] = 3366 s
	σ^2 [E] = 85,447	σ^2 [E] = 52,089
	μ [E] = 32 s	μ [E] = 2 s
Line C	σ^2 [Y] = 126,295	σ^2 [Y] = 73,113
	μ [Y] = 2493 s	μ [Y] = 2352 s
	σ^2 [E] = 56,715	σ^2 [E] = 57,240
	μ [E] = 32 s	μ [E] = 35 s

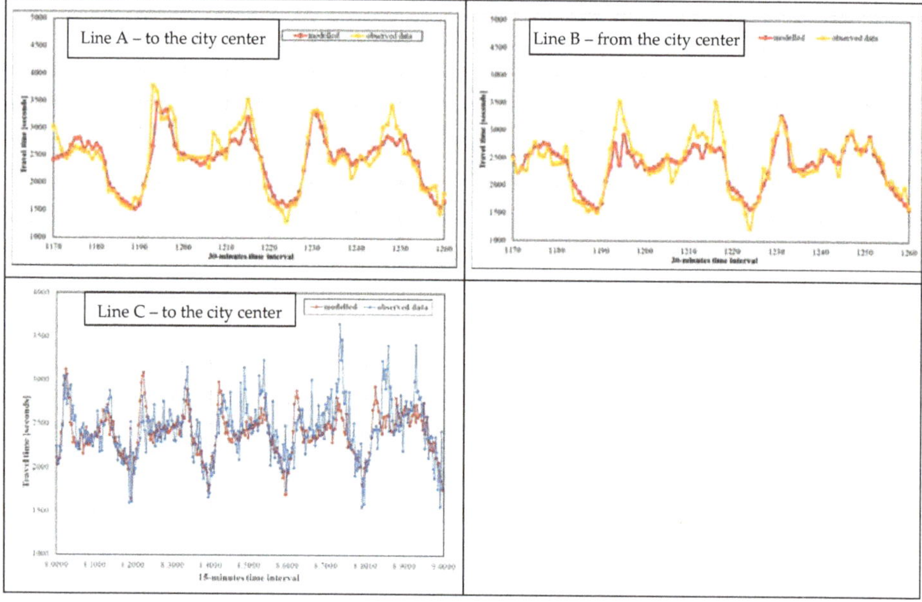

Figure 7. Example of a comparison between observed and forecasted/modelled values.

Table 3. Accuracy indicators of a route of Lines A, B and C.

Line A—Direction to City Centre	Line B—Direction from City Centre	Line C—Direction to City Centre
average e_i = −30.2 s	average e_i = 15.3 s	average e_i = 53.7 s
standard deviation of e_i = 288.7	standard deviation of e_i = 228.2	standard deviation of*ei* = 254.9
MAE = 187.4; RMSE = 290.3; MAPE = 7.4%	MAE = 139.8; RMSE = 254.0; MAPE = 4.2%	MAE = 191.7; RMSE = 260.2; MAPE = 7.9%

4.3. Discussion

The comparison of the bus lines operating in two different (both in size and economic structure) cities pointed out a number of major findings. Bus lines in Lviv operate within the historical centre of the city, while in Rome, the study area merges suburbs with the inner-city area.

As regards spatial form, the inner area of Rome is surrounded by radially distributed roads. Lviv is an Eastern European city with a strong influence from Western cities in its urbanisation. Both the study areas are characterised by narrow streets and buses shares the lanes; therefore, their commercial speed is highly influenced by private traffic. In relation to bus travel time, there are no significant differences in patterns, with high values during morning and afternoon peak hours.

An in-depth comparison between Lviv and Rome showed that the travel time pattern is not strictly dependent on the size of the cities, but it is related to traffic jams (due mainly to lifestyle, e.g., starting of working day). Lviv presents a not-high perturbation during the morning, as indeed happens in Rome, where the high concentration of traffic dominates. Therefore, there is also a very similar pattern in relation to time distribution along the day, although Rome sees activity in the early morning and afternoon, and Lviv does not. The same is reflected in departures from the centre in the afternoon.

Finally, the differences between Rome and Lviv are primarily due to city morphology and lifestyle. The Rome study area has a road network with narrow streets, and it favours an increase of time during the morning peak hour due to high demand, with a high concentration of trip arrivals on time at the workplace or school. The buses share the lanes and necessarily require travel time to be longer. The second reason is lifestyle and the start of work time, in particular of offices and the opening time of shops and stores. While Rome has two main peak hours and more specific regulations (in Rome, a limited traffic zone has been implemented), in Lviv, the circulation of traffic is always allowed in the morning.

According to these first results, the proposed forecasting methods present high performance in covering variances. Of course, the transferability of results in terms of accuracy issues is not direct, and city-specific surveys are needed. Additionally, the general conclusion is that a more systemic approach could be used in forecasting travel time by trying to define more comprehensive methods that are able to capture the characteristics of traffic, taking into account that some features are city-specific. Results, such as those derived in this paper, can contribute to more effective and rational management of resources, offering to mobility agency technicians an easy-to-apply method for designing or revising timetables and scheduled services.

Further analyses are also in progress to improve these first results by developing other analyses through the inclusion of zonal and level-of-service attributes (e.g., traffic volumes, passive and active accessibility) and the characteristics of users that reach these areas for daily activity, as well as alighting and boarding flows. Currently, the travel time variable has been the only independent variable taken into consideration. It does not capture the influence of factors such as weather conditions, road pavement conditions, and pedestrian flows. A time series-based regression model that also includes more independent variables that capture the above effects is under development. Finally, the degree of transferability of the models and the obtainable accuracy is a work in progress.

5. Conclusions

An investigation of bus travel time variability was carried out, and the first results for developing an easy-to-apply methodology for the urban areas of Rome and Lviv are presented. The findings can be used by urban mobility agency technicians for designing transit service timetables and vehicle scheduling in the replanning of existing lines. The developed approach focuses on the systematic components of travel time. The results are mainly devoted to bus lines that share the road lanes with other traffic components (e.g., cars, freight vehicles, and so on). The analyses were performed through time-series methods, which allowed factors such as trends, seasonal variations, cycles, and irregular components to be pointed out. Time of the day as well as the day of the week (i.e., Wednesday vs. Thursday) were discovered to have significant effects on travel time variability, and, if rightly modelled, performing forecasts can be obtained. Moreover, the findings can be integrated into a short-term approach, which, traditionally, is not considered. In fact, the development of models for short-term travel time forecasts that consider the variety of different factors, such as demand, transport conditions, and weather conditions, is a research challenge. From the above statements, the further development of this study germinates the availability of further data on bus line operations and loads, data on other vehicles with whom lanes are shared (e.g., floating car data for revealing local traffic patterns), in-depth analysis of residuals coming from time-series decomposition, as well as an approach to incorporate covariates such as weather data to model travel time as this factor might impact bus line travel time significantly. Finally, bus travel time forecasts are based on profile similarities through the time-series and regression methods.

Author Contributions: Conceptualization, A.C.; methodology, A.C.; software, A.P.; validation, A.C. and A.P.; formal analysis, A.C. and A.P.; investigation, A.C. and A.P.; data curation, A.C. and A.P.; writing—original draft preparation, A.C. and A.P.; writing—review and editing, A.C.; visualization, A.C. and A.P.; supervision, A.C. All authors have read and agreed to the published version of the manuscript.

Funding: This research received no external funding. The APC was funded by MDPI.

Acknowledgments: The authors wish to thank M. Zhuk, V. Kovalyshyn and V. Hilevych, for having shared data from Lviv. The authors wish to thank the anonymous reviewers for their suggestions, which were most useful in revising the paper.

Conflicts of Interest: The authors declare no conflict of interest.

References

1. Russo, F.; Comi, A. Investigating the effects of city logistics measures on the economy of the city. *Sustainability* **2020**, *12*, 1439. [CrossRef]
2. Musolino, G.; Polimeni, A.; Vitetta, A. Freight vehicle routing with reliable link travel times: A method based on network fundamental diagram. *Transp. Lett.* **2018**, *10*, 159–171. [CrossRef]
3. Birr, K.; Jamroz, K.; Kustra, W. Travel time of public transport vehicles estimation. *Transp. Res. Procedia* **2014**, *3*, 359–365. [CrossRef]
4. Fusco, G.; Colombaroni, C.; Isaenko, N. Short-term speed predictions exploiting big data on large urban road networks. *Transp. Res. Part C Emerg. Technol.* **2016**, *73*, 183–201. [CrossRef]
5. Comi, A.; Nuzzolo, A.; Brinchi, S.; Verghini, R. Bus travel time variability: Some experimental evidences. *Transp. Res. Procedia* **2017**, *27*, 101–108. [CrossRef]
6. Karami, Z.; Kashef, R. Smart transportation planning: Data, models, and algorithms. *Transp. Eng.* **2020**, *2*, 100013. [CrossRef]
7. Cats, O. Determinants of bus riding time deviations: Relationship between driving patterns and transit performance. *J. Transp. Eng. Part A Syst.* **2019**, *145*, 04018078. [CrossRef]
8. Cristóbal, T.; Padrón, G.; Quesada-Arencibia, A.; Alayón, F.; De Blasio, G.; Garcia, C.D. Bus travel time prediction model based on profile similarity. *Sensors* **2019**, *19*, 2869. [CrossRef]
9. Wu, J.; Wu, Q.; Shen, J.; Cai, C. Towards attention-based convolutional long short-term memory for travel time prediction of bus journeys. *Sensors* **2020**, *20*, 3354. [CrossRef]
10. Balasubramanian, P.; Rao, K. An adaptive long-term bus arrival time prediction model with cyclic variations. *J. Public Transp.* **2015**, *18*, 6. [CrossRef]
11. Jeong, R.H.; Rilett, L.R. Prediction model of bus arrival time for real-time applications. *Transp. Res. Rec.* **2005**, *1927*, 195–204. [CrossRef]
12. Comi, A.; Nuzzolo, A.; Brinchi, S.; Verghini, R. Bus dispatching irregularity and travel time dispersion. In Proceedings of the 5th IEEE International Conference on Models and Technologies for Intelligent Transportation Systems (MT-ITS), Naples, Italy, 26 June 2017; pp. 856–860.
13. Jeffery, D. *Guidelines for Assessing the Transferability of an Innovative Urban Transport Concept*; NICHES+, Coordination Action funded by the European Commission under the Seventh Framework Programme for R&D, Sustainable Surface Transport; 2011. Available online: www.rupprecht-consult.eu/uploads/tx_rupprecht/NICHES_Guideline_Transferability_01.pdf (accessed on 28 July 2020).
14. Wu, F.; Hu, X.; An, S.; Zhang, D. Exploring passengers' travel behaviors based on elaboration likelihood model under the impact of intelligent bus information. *J. Adv. Transp. Vol.* **2019**, 9095279. [CrossRef]
15. Rupprecht Consult. Guidelines for Developing and Implementing a Sustainable Urban Mobility Plan, Second Edition. 2019. Available online: https://www.eltis.org/mobility-plans/sump-guidelines (accessed on 20 July 2020).
16. Mazloumi, E.; Rose, G.; Currie, G.; Sarvi, M. An integrated framework to predict bus travel time and its variability using traffic flow data. *J. Intell. Transp. Syst.* **2011**, *15*, 75–90. [CrossRef]
17. Kimpel, T.J.; Strathman, J.G.; Callas, S. Improving scheduling through monitoring using AVL/APC data. In Proceedings of the 9th International Conference on Computer-Aided Scheduling of Public Transport (CASPT), San Diego, CA, USA, 9–11 August 2004.
18. Turochy, R.; Smith, B. Measuring variability in traffic conditions by using archived traffic data. *Transp. Res. Rec.* **2002**, *1804*, 168–172. [CrossRef]
19. Ceder, A. *Public Transit Planning and Operation: Modeling, Practice and Behavior*, 2nd ed.; CRC Press: Boca Raton, FL, USA; Taylor & Francis Group: Abingdon, UK, 2015.
20. Bates, J.; Polak, J.; Jones, P.; Cook, A. The valuation of reliability for personal travel. *Transp. Res. Part E* **2001**, *37*, 191–229. [CrossRef]
21. Lam, T.C.; Small, K.A. The value of time and reliability: Measurement from a value pricing experiment. *Transp. Res. Part E* **2001**, *37*, 231–251. [CrossRef]
22. Nuzzolo, A.; Comi, A. A subjective optimal strategy for transit simulation models. *J. Adv. Transp.* **2018**, *2018*, 8797328. [CrossRef]

23. Sun, C.; Arr, G.; Ramachandran, R.P. Vehicle reidentification as method for deriving travel time and travel time distributions. *Transp. Res. Rec.* **2003**, *1826*, 25–31. [CrossRef]
24. Sun, D.; Luo, H.; Fu, L.; Liu, W.; Liao, X.; Zhao, M. Predicting bus arrival time on the basis of global positioning system data. *Transp. Res. Rec.* **2007**, *2034*, 62–72. [CrossRef]
25. Comi, A.; Zhuk, M.; Kovalyshyn, V.; Hilevych, V. Investigating bus travel time and predictive models: A time series-based approach. *Transp. Res. Procedia* **2020**, *45*, 692–699. [CrossRef]
26. Nuzzolo, A.; Comi, A. Advanced public transport and intelligent transport systems: New modelling challenges. *Transp. A Transp. Sci.* **2016**, *12*, 674–699. [CrossRef]
27. Gong, X.; Guo, X.; Dou, X.; Lu, L. Bus travel time deviation analysis using automatic vehicle location data and structural equation modeling. *J. Adv. Transp. Vol.* **2015**. [CrossRef]
28. Kieu, L.M.; Bhaskar, A.; Chung, E. Benefits and issues for bus travel time estimation and prediction. In Proceedings of the Australasian Transport Research Forum 2012, Perth, Australia, 26–28 September 2012.
29. Liu, H. Travel Time Prediction for Urban Networks. Ph.D. Thesis, Netherlands Research School for Transport, Infrastructure and Logistics, TU Delft, The Netherlands, 2008.
30. van Lint, J.W.C.; Hoogendoorn, S.P.; van Zuylen, H.J. Accurate freeway travel time prediction with state-space neural networks under missing data. *Transp. Res. Part C Emerg. Technol.* **2005**, *13*, 347–369. [CrossRef]
31. Hyndman, R.J.; Athanasopoulos, G. *Forecasting: Principles and Practice*, 2nd ed.; OTexts: Melbourne, Australia, 2018. Available online: OTexts.com/fpp2 (accessed on 17 July 2020).
32. Rajbhandari, R. Bus Arrival Time Prediction Using Stochastic Time Series and Markov Chains. Ph.D. Thesis, Faculty of New Jersey Institute of Technology, Newark, NJ, USA, 2006.
33. Bolshinsky, E.; Freidman, R. *Traffic Flow Forecast Survey*; Technical Report; Technion–Israel Institute of Technology: Haifa, Israel, 2012.
34. Fan, W.; Gurmu, Z. Dynamic travel time prediction models for buses using only GPS data. *Int. J. Transp. Sci. Technol.* **2015**, *4*, 353–366. [CrossRef]
35. Moreira-Matias, L.; Mendes-Moreira, J.; de Sousa, J.F.; Gama, J. Improving mass transit operations by using AVL-based systems: A survey. *IEEE Trans. Intell. Transp. Syst.* **2015**, *16*, 1636–1653. [CrossRef]
36. Yu, Z.; Wood, J.S.; Gayah, V.V. Using survival models to estimate bus travel times and associated uncertainties. *Transp. Res. Part C Emerg. Technol.* **2017**, *74*, 366–382. [CrossRef]
37. He, P.; Jiang, G.; Lam, S.K.; Tang, D. Travel-time prediction of bus journey with multiple bus trips. *IEEE Trans. Intell. Transp. Syst.* **2019**, *20*, 4192–4205. [CrossRef]
38. Chen, M.; Liu, X.; Xia, J.; Chien, S.I. A dynamic bus-arrival time prediction model based on APC data. *Comput. Aided Civ. Infrastruct. Eng.* **2004**, *19*, 364–376. [CrossRef]
39. Mendes-Moreira, J.; Jorge, A.M.; de Sousa, J.F.; Soares, C. Comparing state-of-the-art regression methods for long term travel time prediction. *Intell. Data Anal.* **2012**, *16*, 427–449. [CrossRef]
40. Moreira-Matias, L.; Cats, O.; Gama, J.; Mendes-Moreira, J.; de Sousa, J.F. An online learning approach to eliminate bus bunching in real-time. *Appl. Soft Comput.* **2016**, *47*, 460–482. [CrossRef]
41. Williams, B.M.; Hoel, L.A. Modeling and forecasting vehicular traffic flow as a seasonal ARIMA process: Theoretical basis and empirical results. *J. Transp. Eng.* **2013**, *129*, 664–672. [CrossRef]
42. Jeong, R.H. The Prediction of Bus Arrival Time Using Automatic Vehicle Location Systems Data. Ph.D. Thesis, Texas A&M University, College Station, TX, USA, 2004.
43. Billings, D.; Yang, J.-S. Application of the ARIMA models to urban roadway travel time prediction—A case study. In Proceedings of the 2006 IEEE International Conference on Systems, Man and Cybernetics, Taipei, Taiwan, 8 October 2006; Volume 3, pp. 2529–2534.
44. Suwardo, M.N.; Ibrahim, K. ARIMA models for bus travel time prediction. *J. Inst. Eng.* **2010**, *71*, 49–58.
45. Yetiskul, E.; Senbil, M. Public bus transit travel-time variability in Ankara (Turkey). *Transp. Policy* **2012**, *23*, 50–59. [CrossRef]
46. Ahmed, M.; Cook, A. Analysis of freeway traffic time series data by using Box–Jenkins techniques. *Transp. Res. Board* **1979**, *722*, 1–9.
47. Smith, B.L.; Williams, B.M.; Oswald, R.K. Comparison of parametric and nonparametric models for traffic flow forecasting. *Transp. Res. Part C* **2002**, *10*, 303–321. [CrossRef]
48. Lin, W.H.; Zeng, J. Experimental study of real-time bus arrival time prediction with GPS data. *Transp. Res. Rec.* **1999**, *1666*, 101–109. [CrossRef]

49. Kalaputapu, R.; Demetsky, M.J. Modeling schedule deviations of buses using automatic vehicle-location data and artificial neural networks. *Transp. Res. Rec.* **1995**, *1479*, 44–52.
50. Smith, B.L.; Demetsky, M.J. Short-term flow prediction: Neural network approach. *Transp. Res. Rec.* **1995**, *1453*, 98–104.
51. Park, D.; Rilett, L.R. Forecasting freeway link travel times with a multilayer feedforward neural network. *Comput. Aided Civ. Infrastruct. Eng.* **1999**, *14*, 357–367. [CrossRef]
52. Chien, S.; Ding, Y.; Wei, C. Dynamic bus arrival time prediction with artificial neural network. *ASCE J. Transp. Eng.* **2002**, *128*, 429–438. [CrossRef]
53. Chen, M.; Chien, S. Dynamic freeway travel-time prediction with probe vehicle data: Link based versus path based. *Transp. Res. Rec.* **2001**, *1879*, 89–98. [CrossRef]
54. Cathey, F.; Dailey, D. A prescription for transit arrival/departure prediction using AVL data. *Transp. Res. Part C* **2003**, *11*, 241–264. [CrossRef]
55. Shalaby, A.; Farhan, A. Prediction model of bus arrival and departure times using AVL and APC data. *J. Public Transp.* **2004**, *7*, 41–61. [CrossRef]
56. Davis, G.; Nihan, N. Nonparametric regression and short-term freeway traffic forecasting. *J. Transp. Eng.* **1991**, *117*, 178–188. [CrossRef]
57. Sun, H.; Liu, H.X.; Xiao, H.; He, R.R.; Ran, B. Use of local linear regression model for short-term traffic forecasting. *Transp. Res. Rec.* **2003**, *1936*, 143–150. [CrossRef]
58. Qi, Y.; Smith, B.L. Identifying nearest-neighbors in a large-scale incident data archive. *Transp. Res. Rec.* **2004**, *1879*, 89–98. [CrossRef]
59. Zhang, X.; Chen, M. Quantifying the impact of weather events on travel time and reliability. *J. Adv. Transp. Vol.* **2019**. [CrossRef]
60. Cleveland, R.B.; Cleveland, W.S.; Terpenning, I. STL: A seasonal-trend decomposition procedure based on loess. *J. Off. Stat.* **1990**, *6*, 3.
61. Jeon, S.; Hong, B. Monte Carlo simulation-based traffic speed forecasting using historical big data. *Future Gener. Comput. Syst.* **2016**, *65*, 182–195. [CrossRef]
62. Bai, C.; Peng, Z.R.; Lu, Q.C.; Sun, J. Dynamic bus travel time prediction models on road with multiple bus routes. *J. Adv. Transp. Vol.* **2015**. [CrossRef]

Article

Cost Estimating Using a New Learning Curve Theory for Non-Constant Production Rates

Dakotah Hogan [1], John Elshaw [2,*], Clay Koschnick [2], Jonathan Ritschel [2], Adedeji Badiru [3] and Shawn Valentine [4]

1 Air Force Cost Analysis Agency, Deputy Assistant Secretary for Cost and Economics, Joint Base Andrews, MD 20762, USA; dakotah.hogan.1@us.af.mil
2 Department of Systems Engineering & Management, Air Force Institute of Technology, Wright-Patterson AFB, OH 45433, USA; clay.koschnick@afit.edu (C.K.); jonathan.ritschel@afit.edu (J.R.)
3 Graduate School of Engineering and Management, Air Force Institute of Technology, Wright-Patterson AFB, OH 45433, USA; adedeji.badiru@afit.edu
4 Estimating Research & Technology Advising Branch, Cost and Economics Division, Air Force Lifecycle Management Center, Wright-Patterson AFB, OH 45433, USA; shawn.valentine@us.af.mil
* Correspondence: john.elshaw@afit.edu; Tel.: +1-937-255-3636 (ext. 4650)

Received: 27 August 2020; Accepted: 9 October 2020; Published: 16 October 2020

Abstract: Traditional learning curve theory assumes a constant learning rate regardless of the number of units produced. However, a collection of theoretical and empirical evidence indicates that learning rates decrease as more units are produced in some cases. These diminishing learning rates cause traditional learning curves to underestimate required resources, potentially resulting in cost overruns. A diminishing learning rate model, namely Boone's learning curve, was recently developed to model this phenomenon. This research confirms that Boone's learning curve systematically reduced error in modeling observed learning curves using production data from 169 Department of Defense end-items. However, high amounts of variability in error reduction precluded concluding the degree to which Boone's learning curve reduced error on average. This research further justifies the necessity of a diminishing learning rate forecasting model and assesses a potential solution to model diminishing learning rates.

Keywords: learning curve; forecasting; production cost; cost estimating

---

1. Introduction

The U.S. Government Accountability Office (GAO) critiqued the cost and schedule performance of the Department of Defense (DoD)'s $1.7 trillion portfolio of 86 major weapons systems in their 2018 "Weapons System Annual Assessment." The GAO cited realistic cost estimates as a reason for the relatively low cost growth of the portfolio in comparison to earlier portfolios [1]. Congress and its oversight committees maintain a watchful eye on the DoD's complex and expensive weapons system portfolio. Inefficient programs are scrutinized and may be terminated if inefficiencies persist. Funding of inefficient programs will also lead to the underfunding of other programs. In the public sector, these terminated and underfunded programs may result in capability gaps that negatively impact our nation's defense. In the private sector, the inefficient use of resources often spells failure for a company.

A key to the efficient use of resources is accurately estimating the resources required to produce an end-item. Learning curves are a popular method of forecasting required resources as they predict end-item costs using the item's sequential unit number in the production line. Learning curves are especially useful when estimating the required resources for complex products. The most popular learning curve models used in the government sector are over 80 years old and may be outdated in today's technology-rich production environment. Additionally, researchers have demonstrated both theoretically and empirically that the effects of learning slow or cease over time [2–4].

A new model, named Boone's learning curve, has been recently proposed to account for diminishing rates of learning as more units are produced [5]. The purpose of this research is to survey the need for alternative learning curve models and further examine how Boone's learning curve performs in comparison to the traditional learning curve theories in predicting required resources. This research uses a large number of diverse production items to compare Boone's model to the traditional theories of Wright and Crawford. While many different learning curve models exist (i.e., DeJong, Stanford B, Sigmoid, etc.), some of these others may not be as accurate in cases where the learning rate decreases over time. The next section is a review of the learning curve literature relevant to diminishing learning rates, followed by a description of our methodology and analysis to compare Boone's learning curve to traditional models. We conclude the paper discussing managerial implications and limitations followed by recommendations for the way forward.

2. Literature Review and Background

The two learning curve models cited by the GAO Cost Estimating and Assessment Guide (2009) are Wright's cumulative average learning curve theory developed in 1936 and Crawford's unit learning curve theory developed in 1947. Although both learning curve theories use the same general equation, the theories have contrasting variable definitions. Wright's learning curve is shown in Equation (1):

$$\overline{Y} = Ax^b \tag{1}$$

where $\overline{Y}$ is the cumulative average cost of the first x units, A is the theoretical cost to produce the first unit, x is the cumulative number of units produced, and b is the natural logarithm of the learning curve slope (LCS) divided by the natural logarithm of two. Note, the LCS is the complement of the percent decrease in cost as the number of units produced doubles. For example, with a learning curve slope of 80% and a first unit cost of 100 labor hours, the average cost of the first two units would be 80 labor hours, or 60 labor hours for the second unit. Regardless of the number of units produced, there is a constant decrease in labor costs with each doubling of units due to the constant learning rate.

Several years following the creation of Wright's cumulative average learning curve theory, J.R. Crawford formulated the unit learning curve theory. Crawford's theory deviates from Wright's by assuming that the individual unit cost (as opposed the cumulative average unit cost) decreases by a constant percentage as the number of units produced doubles. Crawford's model is shown in Equation (2):

$$Y = Ax^b \tag{2}$$

where Y is the individual cost of unit x, A is the theoretical cost of the first unit, x is the unit number of the unit cost being forecasted, and b is the natural logarithm of the LCS divided by the natural logarithm of two. For example, with a learning curve slope of 80% and a first unit cost of 100 labor hours, the cost of the second unit would be 80 labor hours. Note, Crawford's unit theory is the similar to Wright's in function form; but the difference arises in the variable interpretation lead to a different forecast.

Figure 1 below shows a comparison between Wright's and Crawford's theories using the two numerical examples provided. Cumulative average theory and unit theory will produce different predicted costs provided the same set of data despite all predicted costs being normalized to unit costs. Figure 1 demonstrates this point where unit theory was used to generate data using a first unit cost of 100 and a learning curve slope of 90%. The original unit theory data was converted to cumulative averages in order to estimate cumulative average theory learning curve parameters.

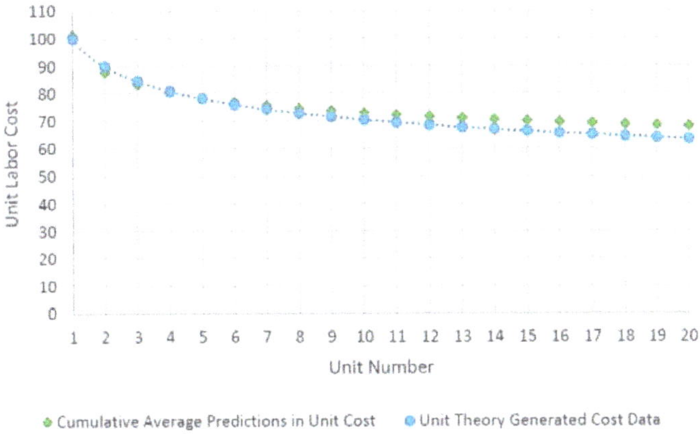

Figure 1. Wright's Cumulative Average Theory vs. Crawford's Unit Theory.

Cumulative average theory learning curve parameters. Cumulative average theory estimated a learning curve slope of 93% and a first unit cost of 101.24. These Cumulative Average Theory parameters were then used to predict cumulative average costs. These predicted costs were then converted to unit costs. This conversion allows for the cumulative average predictions to be directly compared to the original Unit Theory generated data. As shown in Figure 1, the cumulative average learning curve predictions first overestimate, then underestimate, and ultimately overestimate the generated unit theory data for all remaining units. Together, Wright's and Crawford's theories form the basis of the traditional learning curve theory.

One assumption of these traditional learning curve theories is that they only apply to processes that may benefit from learning. Typically, these costs are only a subset of total program costs; hence appropriate costs must be considered when applying learning curve theory to yield viable parameter estimates. In a complex program, costs can be viewed in a variety of ways to include recurring and non-recurring costs, direct and indirect costs, and costs for various activities and combinations of end-items that can be stated in units of hours or dollars. Learning curve analysis focuses solely on recurring costs in estimating parameters because these costs are incurred repeatedly for each unit produced [6]. Researchers have also focused solely on direct labor costs due to the theoretical underpinnings of learning occurring at the laborer level [2,3]. Additionally, researchers have historically studied end-items that include only the manufactured or assembled hardware and software elements of the end-item [2,3]. Lastly, labor hours in lieu of labor dollars are generally used in analysis so that data can be compared across fiscal years without the need to adjust for inflation. Therefore, the literature indicates using direct, recurring, labor costs in units of labor hours. These costs should be considered only for the certain elements that include the manufacturing or assembly of hardware and software of an end-item.

An implicit assumption in the traditional learning curve theories is that knowledge obtained through learning does not depreciate. However, empirical evidence demonstrates that knowledge depreciates in organizations [7,8]. Argote [7] showed that knowledge depreciation occurs at both the individual and the organizational levels. Many variations of the traditional models make use of the concept of performance decay (commonly called forgetting) to model non-constant rates of learning. Forgetting and its relationship to learning can take many forms and is essential to consider in contemporary learning curve analysis.

Forgetting is the concept that an individual or organization will experience a decline in performance over time resulting in non-constant rates of learning. Badiru [4] theorizes that forgetting and resulting performance decay is a result of factors "including lack of training, reduced retention of skills, lapse in performance, extended breaks in practice, and natural forgetting" (p. 287). According to Badiru [4], these factors may be caused by internal processes or external factors. Badiru [4] lists three cases in which forgetting arises. First, forgetting may occur continuously as a worker or organization progresses down the learning curve due in part to natural forgetting [4]. The impact of forgetting may not wholly eclipse the impact of learning but will hamper the learning rate while performance continues to increase at a slower rate. Second, forgetting may occur at distinct and bounded intervals, such as during a scheduled production break [4] or towards the end of production as workers are transferred to other duties. Finally, forgetting may intermittently occur at random times and for stochastic intervals such as during times of employee turnover [4]. Others have expanded on the causes of forgetting and have drawn similar conclusions to Badiru [4,9–11]. This decline in performance decays the learning rate and causes longer manufacturing times and higher costs than would be forecasted using traditional learning curve theory.

The concept of forgetting and its impact on non-constant rates of learning has proven relevant in contemporary learning curve research. Several forgetting models have been developed to include the learn-forget curve model (LFCM) [11], the recency model (RCM) [12], the power integration and diffusion (PID) model [13], and the Depletion-Power-Integration-Latency (DPIL) model [13] among others [10]. However, these forgetting models focus solely on the phenomenon of forgetting due to interruptions of the production process [9,10,14]. Jaber [9] states that "there has been no model developed for industrial settings that considers forgetting as a result of factors other than production breaks" (pp. 30–31) and mentions this as a potential area of future research. Although forgetting models have emerged after Jaber's [9] article, a review of the popular forgetting models cited confirms Jaber's statement.

A related concept to the forgetting phenomenon is the plateauing phenomenon. According to Jaber [9] (2006), plateauing occurs when the learning process ceases and manufacturing enters a production steady state. This ceasing of learning results in a flattening or partial flattening of the learning curve corresponding to rates of learning at or near zero. There remains debate as to when plateauing occurs in the production process or if learning ever ceases completely [3,9,15–17]. Jaber [9] provides several explanations to describe the plateauing phenomenon that include concepts related to forgetting. Baloff [18,19] recognized that plateauing is more likely to occur when capital is used in the production process as opposed to labor. According to some researchers, plateauing can be explained by either having to process the efficiencies learned before making additional improvements along the learning curve or to forgetting altogether [20]. According to other researchers, plateauing can be caused by labor ceasing to learn or management's unwillingness to invest in capital to foster induced learning [21]. Related to this underinvestment to foster induced learning, management's doubt as to whether learning efficiencies related to learning can occur is cited as another hindrance to constant rates of learning [22]. Li and Rajagopalan [23] investigated these explanations and concluded that no empirical evidence supports or contradicts them while ascribing plateauing to depreciation in knowledge or forgetting. Jaber [9] concludes that "there is no tangible consensus among researchers as to what causes learning curves to plateau" and alludes that this is a topic for future research (pp. 30–39).

Despite the controversy in the research surrounding forgetting and plateauing effects, empirical studies have shown learning curves to exhibit diminishing rates of learning. For instance, the plateauing phenomenon at the tail end of production was investigated by Harold Asher in a 1956 RAND study. The U.S. Air Force contracted RAND after the service noticed traditional learning curves were underestimating labor costs at the tail end of production [3]. Asher intended to study if the logarithmically transformed traditional learning curves were approximately linear. This linearity would indicate constant rates of learning throughout the production cycle. The alternative hypothesis for these learning curves was a convexity of the logarithmically-transformed traditional learning curves that would indicate diminishing rates of learning as the number of units increased [3]. An example of a learning curve with a diminishing learning rate is shown in Figure 2 in logarithmic scale. The first unit cost is 100 with an initial learning curve slope of 80% decaying at a rate of 0.25% with each additional unit. For example, the second unit's learning curve slope is 80.25%.

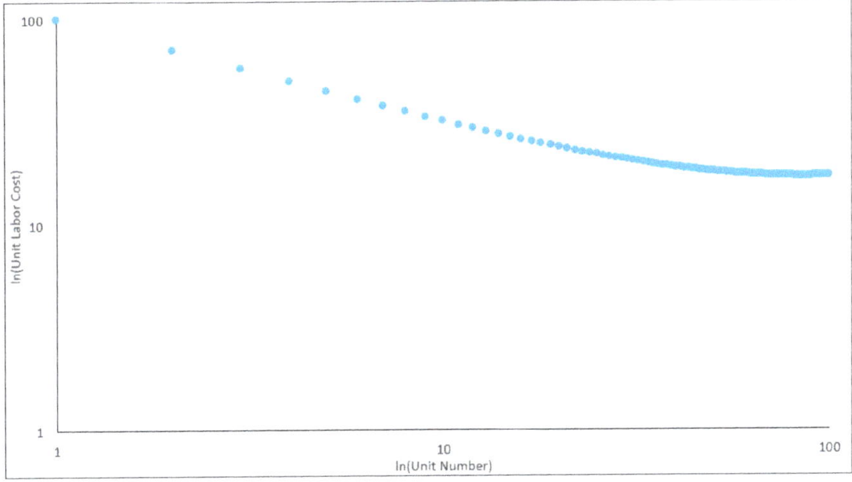

Figure 2. Unit Theory learning curve with a Decaying Learning Curve Slope.

Asher investigated this hypothesis of convex logarithmically transformed learning curves by analyzing the learning curves of the various shops within a manufacturing department producing aircraft. Asher used airframe cost data with the appropriate amount of detail to perform a learning curve analysis on the lower level job shops within the manufacturing department. He divided the eleven major kinds of aircraft manufacturing operations into four shop groups each with a set of direct labor cost data [3]. If non-constant rates of learning were present, the shop group curves would differ in their rates of learning and may themselves be convex in logarithmic scale. This would indicate their aggregate learning curve would also be convex in logarithmic scale.

Asher's results showed that the learning curves of the manufacturing shop group had different learning slopes and were convex in logarithmic scale [3]. Asher claims the convexity within the manufacturing shop group learning curves is due to the disparate operations within the job shops and stated that each had their own unique learning curve [3]. He asserts that a linear approximation is reasonable for a relatively small quantity of airframes produced but becomes increasingly unwarranted for larger quantities. This is due in part because larger quantities of produced end-items are likely to experience diminishing rates of learning. Moreover, highly aggregated learning curves are also likely to experience diminishing rates of learning. Because the aggregated manufacturing cost curve is usually the lowest level of detail on which learning curve analysis is performed, the manufacturing cost curve will have diminishing rates of learning as cumulative output increases. These results further justify a learning curve model with diminishing rates of learning.

Wright's and Crawford's learning curve theories provided the basis of the traditional approach that learning occurs at a constant rate as the number of units produced increases. Since this initial discovery, several log-linear learning curve models were founded in attempts to more accurately model data from manufacturing processes. These contemporary models diverge from constant rates of learning by including adjustments in various forms. The six most popular models (including the traditional model) are shown in Figure 3 in logarithmic scale and include log-log graphing lines to more clearly illustrate the differences between models. These illustrated models include the traditional log-linear model or Wright/Crawford curves, the plateau model [19], the Stanford-B model [24], the De Jong model [25], the S-curve model [21], and Knecht's upturn model [26].

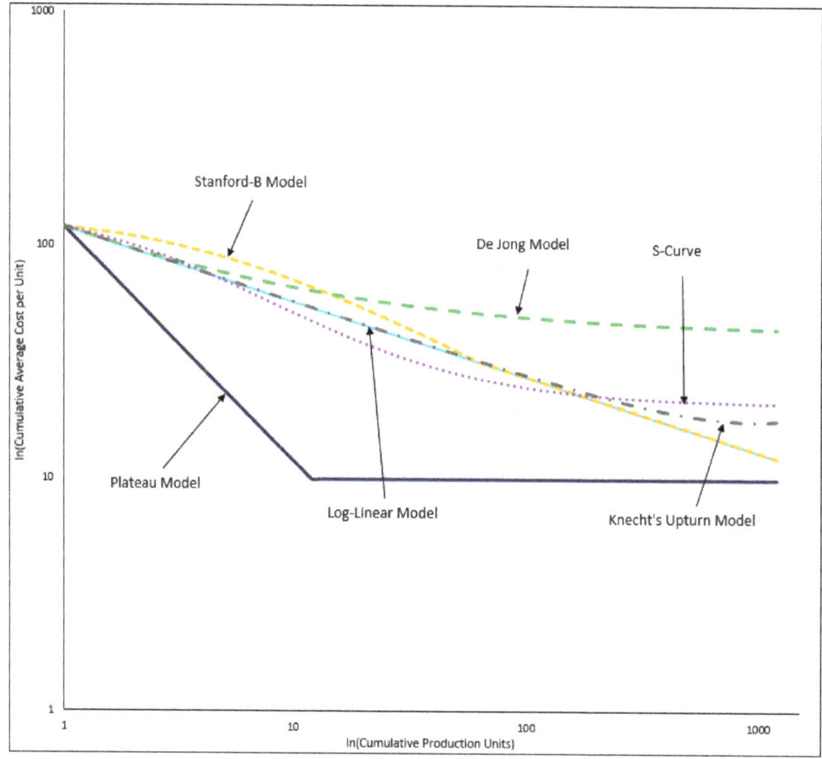

Figure 3. Comparison of Learning Curve Models (adapted from Badiru [27]).

Recent studies have investigated whether the Stanford-B, De Jong, and S-Curve models more accurately predict program costs in comparison to the traditional theories. Moore [16] and Honious [17] studied how prior experience in the manufacturing of an end-item along with the proportion of touch labor in the manufacturing process affected the accuracy of the Stanford-B, De Jong, and S-curve models in comparison to the traditional models. The authors concluded that these models improved upon the traditional curves for only a narrow range of parameter values. Their research provided insight that the traditional learning curve models become less accurate at the tail-end of production when the proportion of human labor is high in the manufacturing process. Moreover, Honious [17] explicitly references a plateauing effect at the end of production. These findings provide further justification for investigating non-constant rates of learning.

The Stanford-B, De Jong, and S-Curve univariate models illustrated in Figure 3 alter the resulting learning curve slope based on alterations to the theoretical first unit cost parameter A. However, the learning curve slopes of these models are not directly a function of the number of cumulative units produced. The plateau model and Knecht's upturn model also illustrated in Figure 3 each produce a learning curve whose slope is directly affected by the number of cumulative units produced. The plateau model uses a step function to reduce the learning rate to 0% (i.e., the learning curve slope is 100%) past a certain number of cumulative units produced. In contrast, Knecht's Upturn Model amends the learning curve exponent term b by multiplying b by Euler's number e raised to the term of a constant multiplied by the number of cumulative units produced. Mathematically, this is expressed $\overline{Y} = Ax^{b \cdot e^{xc}}$, where $\overline{Y}$ is the cumulative average unit cost, A is the theoretical first unit cost, x is the number of cumulative units produced, b is the natural logarithm of the learning curve slope divided by the natural logarithm of 2, and c is a constant. The forgetting models stated within the manuscript also amend the learning curve slope based indirectly on the number of cumulative units but only apply when interruptions to the production process occur.

In response to these researchers' findings, Boone [5] developed a learning curve model with a learning rate that diminishes as more units are produced. Conversely, the traditional learning curve theories diminish the rate of cost reductions as the number of units produced doubles. However, the existing literature provides evidence that the cost reductions with each doubling of units may not be constant as the number of units produced increases. Therefore, Boone [5] sought to attenuate the cost reductions that occur with each doubling of units produced by decreasing the learning rate as the number of units increases.

Boone [5] devised a model that decreases the learning curve exponent b as the number of units produced x increases. He first considered a model without an additional parameter to reduce the learning curve exponent b directly by the unit number. However, he decided to temper the effect each additional unit has on the parameter b by adding an additional parameter c. The resulting learning curve is shown in Equation (3):

$$\overline{Y} = Ax^{\frac{b}{1+\frac{x}{c}}} \tag{3}$$

where $\overline{Y}$ is the cumulative average cost of the first x units, A is the theoretical cost to produce the first unit, x is the cumulative number of units produced, b is the natural logarithm of the learning curve slope (LCS) divided by the natural logarithm of two, and c is a positive decay value. For example, a learning curve slope of 80%, first unit cost of 100 labor hours, and decay value of 100, Boone's model yields a cumulative average cost at the second unit of 80.35 labor hours—or 60.70 labor hours for the second unit. What began as an 80% learning curve model has decayed to an 80.35% learning curve for the second unit. In comparison to Wright's learning curve using the same parameters, the effect of learning has decreased slightly in the production of unit two. The inclusion of the decay value increases the learning curve slope, and hence decreases the learning rate as more units are produced. Note, Boone's model can also be modified to incorporate Crawford's unit theory–refer to Equation (3) for the necessary modifications.

Boone's learning curve diverges from the constant learning assumptions in both Wright's and Crawford's learning curve models by incorporating the unit number in the denominator of the exponent—thus decreasing the effect of b as the number of units produced increases. Furthermore, the decay value moderates this diminishing effect, so the amount of learning decreases more slowly. In general, Boone's model is flatter near the end of production and steeper in the early stages compared to the traditional theories. Note, as the decay value approaches zero (holding other factors constant), the exponent term approaches zero representing a learning curve slope approaching 100%. As the decay value approaches infinity, the parameter b remains constant, and Boone's learning curve simplifies to the traditional learning curve [5].

Boone [5] tested his learning curve using unit theory to provide a consistent comparison to Crawford's learning curve. Based on the scope of his research and lack of comparison using cumulative average theory, a more robust examination and analysis of Boone's learning curve should be accomplished.

3. Methodology

One goal of this research is to examine the accuracy of Boone's learning curve in comparison to the popular Wright and Crawford learning curve theories. In order to perform this analysis, production cost and quantity data from a diverse set of DoD systems was collected from government Functional Cost-Hour Reports, Progress Curve Reports, and the Air Force Life Cycle Management Center Cost Research Library. The dataset consisted of recurring costs (either in dollars or labor hours) by production lot for 169 unique end-items. Our data included end-items from a variety of systems (i.e., bomber, cargo, and fighter aircraft, missiles, and munitions), contractors, and time periods (1957–2018). Additionally, only production runs with at least four lots were included. The dataset for the Cumulative Average Theory analysis only includes 140 of the 169 end-items. This theory relies on continuous data because each lot's cumulative average cost and cumulative quantity is a function of all previous lots' costs and quantities. In order to compare Boone's model to the traditional theories, each model will be fitted to data: (1) Boone's and Wright's models using cumulative average theory, and (2) Boone's and Crawford's models using unit theory. Then, the predicted values for each model will be compared to the actual costs using root mean squared error (RMSE) and mean absolute percentage error (MAPE).

Labor costs were collected from the work breakdown structure (WBS) for the specific item being manufactured (e.g., aircraft frame) or from the documentation provided by the government. Our data included three broad functional cost categories: labor, material, and other. These costs are included in both forms of recurring and non-recurring costs. There are also four functional labor categories delineated that include manufacturing, tooling, engineering, and quality control labor. These four labor category costs, when summed with the material costs and other costs, comprise the total cost for each WBS element for recurring and non-recurring costs.

The definition for the manufacturing labor cost category most clearly aligns with the extant literature to be the focus as the pertinent labor cost category for learning curve research. According to the WBS elements, the manufacturing labor category "includes the effort and costs expended in the fabrication, assembly, integration, and functional testing of a product or end item. It involves all the processes necessary to convert raw materials into finished items [28]." This manufacturing labor category aligns with the categories examined by Wright, which he called "assembly operations [2]," along with those cost categories Crawford studied, which he called "airframe-manufacturing processes [3]." Therefore, the manufacturing labor cost category as defined by the government is associated with the types of labor costs studied by traditional learning curve theorists and succeeding research.

The learning curve parameters for each model (i.e., Equations (1)–(3)) will be estimated by minimizing the sum of squares error (SSE) using Excel's generalized reduced gradient (GRG) nonlinear solver and evolutionary solver. The SSE is calculated by squaring the vertical difference of the observed data and predicted data for each lot and summing these squared differences across all lots.

With lot data, cumulative theory models can be estimated directly. Conversely, when utilizing unit learning curve theory, Crawford's and Boone's models are estimated using an iterative process based on lot midpoints, adapted from Hu and Smith [29]. The algebraic lot midpoint is defined as "the theoretical unit whose cost is equal to the average unit cost for that lot on the learning curve" [6]. The lot midpoint supplants using sequential unit numbers when using lot cost data.

Lot midpoints and model parameters are calculated iteratively due to the lack of a closed-form solution for the lot midpoint. First, an initial lot midpoint (for each lot) is determined using a parameter-free approximation formula [6]—see Equation (4):

$$\text{Lot Midpoint(LMP)} = \frac{F + L + 2\sqrt{FL}}{4} \tag{4}$$

where F is the first unit number in a lot and L is the last unit number in a lot. These lot midpoint estimates are then used to estimate the learning curve parameters for Crawford's model (Equation (2)) using the GRG non-linear optimization algorithm. Next, using the estimated parameter b, a new set of lot midpoints are determined using a simple and popular formula—Asher's Approximation [6]; see Equation (5):

$$\text{Lot Midpoint} \approx \left[\frac{\left(L + \frac{1}{2}\right)^{b+1} - \left(F - \frac{1}{2}\right)^{b+1}}{(L - F + 1)(b + 1)}\right]^{\left(\frac{1}{b}\right)} \tag{5}$$

where F is the first unit number in a lot, L is the last unit number in a lot, and b is the estimated value from Equation (2). Learning curve parameters will then be re-estimated using these more precise lot midpoint estimates. The iterative process is repeated until changes between successive values of the estimated lot midpoints and b are sufficiently small [29] (see Appendix A for a summary of this process). In order to use an iterative process for Boone's model, Asher's Approximation from Equation (5) was adapted to incorporate Boone's decaying learning curve slope. This adaptation allows the lot costs of Boone's learning curve to decrease as more units are produced which affects the lot midpoint estimates; the formula is shown in Equation (6):

$$\text{Lot Midpoint}_{\text{i}} \approx \left[\frac{\left(L+\frac{1}{2}\right)^{b'+1}-\left(F-\frac{1}{2}\right)^{b'+1}}{(L-F+1)(b'+1)}\right]^{\left(\frac{1}{b'}\right)} \tag{6}$$

where F is the first unit number in a lot, L is the last unit number in a lot, $b' = \frac{b}{1+\left(\frac{LMP_{i-1}}{c}\right)}$, and i is the iteration number.

This iterative process of calculating the lot mid-point then solving a non-linear least squares problem requires the execution of a series of non-linear optimization algorithms. Boone's model requires the GRG algorithm which found solutions in a longer but still reasonable amount of time. While more burdensome than the traditional models due to the longer run time and the requirement to provide bounds for the parameters. For Boone's model, the bounds for A and b have a fairly straightforward basis by which to define the bounds. In practice, the A parameter is often supported by a point estimate of the cost of the first theoretical unit. Thus, a bound can be built around this value with tools such as a confidence interval. The b parameter is defined by the learning curve slope which for all practical purposes will be in the (0, 1) interval—most likely on the higher end. As for the c parameter, the basis for the bound is more of a challenge. From a model implementation standpoint, the bound can be arbitrarily large if a long solve time is not limiting. Practically, the bound should be reasonably set; this aspect of the model is an avenue of future research which is discussed in the conclusion. This algorithm does allow the analyst to define stopping conditions such as convergence threshold, maximum number of iterations, or maximum amount of time. Additionally, there is an option called multi-start which uses multiple initial solutions to help locate a global solution verse possibly only finding a local solution. These options allow the user to mitigate the extra burden if necessary. Overall, the computing burden to calculate these models was on the order of minutes per weapon system.

The final estimated parameters for Boone's model and the traditional learning curves were used to create predicted learning curves. These predicted curves were then compared to observed data. Total model error was calculated by comparing the difference between observations and predicted values to understand how accurately the models explained variability in the data. Two measures were used to determine the overall model error. The first error measure was Root Mean Square Error (RMSE) that is calculated by taking the square root of the total SSE divided by the number of lots. RMSE is not robust to outliers—i.e., the effects of outliers may unduly influence this measure. RMSE is often interpreted as the average amount of error of the model as stated in the model's original units.

The second measure was mean absolute percentage error (MAPE). MAPE is calculated by subtracting the predicted value from the observed value, dividing this difference by the observed value, taking the absolute value, and multiplying by 100%. These absolute percent errors are then summed over all observations and divided by the total number of observations. MAPE provides a unit-less measure of accuracy and is interpreted as the average percent of model inaccuracy. Unlike RMSE, MAPE is robust to outliers.

After calculating these measures of overall model error, a series of paired difference t-tests are conducted to determine if reductions in error from Boone's learning curve are statistically significant. In order to conduct the first paired difference t-test, Boone's learning curve RMSE using cumulative average theory will be subtracted from Wright's learning curve RMSE, and the difference will be divided by Wright's learning curve RMSE. This calculation will yield a percentage difference rather than raw difference to compare end-items of varying differences in magnitude equitably. The null hypothesis posits that Boone's learning curve results in an equal amount (or more) of error in predicting observed values compared to Wright's learning curve. The alternative hypothesis is that the percentage difference is greater than zero. Support for the alternative hypothesis signifies that Boone's learning curve results in less error predicting observed values than Wright's learning curve. This methodology will be repeated five times to examine each learning curve theory using the two error measures and the different units of production costs—see Table 1.

Table 1. Paired Difference Hypothesis Tests Conducted.

Learning Curve Theory	Error Measure	Units of Measure
Cumulative Average Theory	Root Mean Squared Error Percentage Difference	Total Dollars(K)
		Labor Hours
	Mean Absolute Percent Error Percentage Difference	Total Dollars(K)&Labor Hours Combined
Unit Theory	Root Mean Squared Error Percentage Difference	Total Dollars(K)
	Mean Absolute Percent Error Percentage Difference	Total Dollars(K)&Labor Hours Combined

An assumption to utilize the paired difference t-test is that the data are approximately normally distributed. For hypothesis tests with large sample sizes, the central limit theorem can be invoked. Alternatively, a Shapiro–Wilk test will be used to evaluate the normality assumption for small samples. If the Shapiro–Wilk test does not support the normality assumption, the non-parametric Wilcoxon Rank Sum test will be used. A 0.05 level of significance will be used for all statistical tests.

4. Analysis & Results

The detailed results for Wright's and Boone's learning curves using cumulative average theory are provided in Appendix B Tables A1 and A2. A total of 118 end-items in units of total dollars and 22 components in units of labor hours were analyzed. Each entry lists the program number, number of production lots, number of items produced, type of end-item, and units of the production costs. Additionally, each entry lists both error measures and the respective percent difference between the models. Positive (negative) differences indicate Boone's model has less (more) error than Wright's.

Boone's curve performs better for two reasons. First, Boone's model can explain costs to at least the same degree of accuracy as the traditional learning curve theories due to the extra parameter. Second, increased accuracy could also be explained by Boone's functional form. Despite these theoretical explanations, Boone's model had more error than Wright's for some observations; these negative percentage differences occur because an upper bound was placed on Boone's decay value. An upper bound of 5000 was used for the decay value (same as Boone's original paper). The practical effect of this particular bound can be observed by the number of end-items where the traditional models significantly outperformed Boone's (i.e., a MAPE difference larger than 0.5%): 7 out of 140 for cumulative average theory and 15 out 169 for unit theory. Thus, the majority of the results were not affected by this artificial limitation which was chosen by trial and error. In practice, the bound could be set arbitrarily large so that it is not binding. Boone's learning curve. This upper bound was necessary s since the GRG algorithm requires bounds on the estimated parameters.

Some percentage error differences are approximately (but not exactly) zero. Observations with percentage error differences of approximately zero were defined as those within the bounds $(-0.25\%, 0.25\%)$. These bounds were used by the researchers to distinguish between observations with approximately zero and non-zero percentage error differences in order to inform the descriptive statistics.

Boone's model had less error for 41% of observations, was approximately equal to Wright's for 50% of observations, and had more error for 9% of observations. While Boone's model is an improvement on Wright's for some observations, many times the models fit the data equally well (i.e., an approximate zero difference).

The results of the paired difference t-tests for cumulative average theory are shown in Table 2 and a sample graph is shown in Figure 4. No outliers, as defined by a value which fell more than three interquartile ranges from the upper 90% and lower 10% quantiles, were present in any of the tests.

Table 2. Cumulative Average Theory Descriptive and Inferential Statistics.

					Hypothesis Test: H0: $\mu \leq 0$ H_A: $\mu > 0$			
Learning Curve Theory	**Error Measure**	**Units of Measure**	**Sample Mean ($\bar{x}$)**	**Sample Standard Deviation (s)**	**Number of Observations**	**Test Statistic**	**p-Value**	**Result**
Cumulative Average Theory	Root Mean Squared Error Percentage Difference	Total Dollars(K)	19.3%	28.90%	118	7.23	<0.001	Reject H0
		Labor Hours	15.20%	31.20%	22	18.5	0.28	Fail to reject H0
	Mean Absolute Percent	Total Dollars(K)&Labor Hours Combined	18.60%	29.50%	140	7.45	<0.001	Reject H0

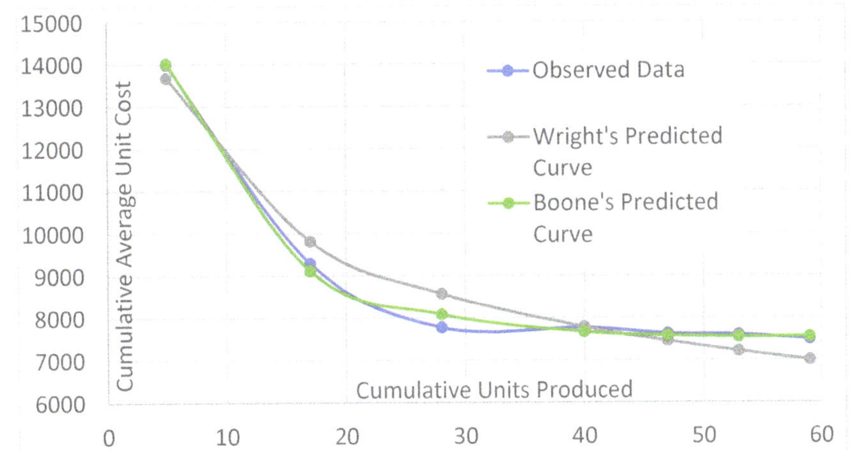

Figure 4. Comparison of Program 20 PME Air Vehicle.

The results of these hypothesis tests were mixed. For the RMSE percentage difference (measured in total dollars) and MAPE percentage difference, the paired difference t-tests led to rejection of the null hypothesis—indicating the increase in accuracy is statistically significant. Conversely, RMSE percentage difference (measured in hours) failed to reject the null hypothesis. Due to the small sample size, large sample theory could not be used, and the data failed a Shapiro–Wilk test (p-value = 0.721). Therefore, a Wilcoxon rank signed test was used. This indicates that Boone's improvement in accuracy over Wright's is not statistically significant when costs are measured in labor hours. However, small sample sizes can cause paired difference tests to have low power that may cause hypothesis tests to incorrectly fail to reject the null hypothesis [30].

Now considering unit theory, the results from Crawford's and Boone's learning curve models are presented in Appendix B. A total of 141 end-items (measured in total dollars) and 28 end-items (measured in labor hours) were analyzed.

Similar to cumulative average theory, observations with percent error differences of approximately zero were defined as those within the bounds $(-0.25\%, 0.25\%)$. Boone's model had less error for 43% of observations across all percent difference error measures in comparison to crawford's learning curve.

Boone's learning curve error was approximately equal for 52% of observations, and had more error for 5% of observations.

The results of the paired difference testing for unit theory are provided in Table 3 and a sample graph is shown in Figure 5. Again, no outliers were present in any of the paired difference *t*-tests.

Table 3. Unit Theory Descriptive and Inferential Statistics.

					Hypothesis Test: H0: $\mu \leq 0$ H_A: $\mu > 0$			
Learning Curve Theory	**Error Measure**	**Units of Measure**	**Sample Mean ($\bar{x}$)**	**Sample Standard Deviation (s)**	**Number of Observations**	**Test Statistic**	***p*-Value**	**Result**
Unit Theory	Root Mean Squared Error Percentage Difference	Total Dollars(K)	13.80%	22.70%	141	7.23	<0.001	Reject H0
		Labor Hours	6.00%	14.80%	28	74.00	0.046	Reject H0
	Mean Absolute Percent Error Percentage Difference	Total Dollars(K) &Labor Hours Combined	11.30%	23.10%	169	6.36	<0.001	Reject H0

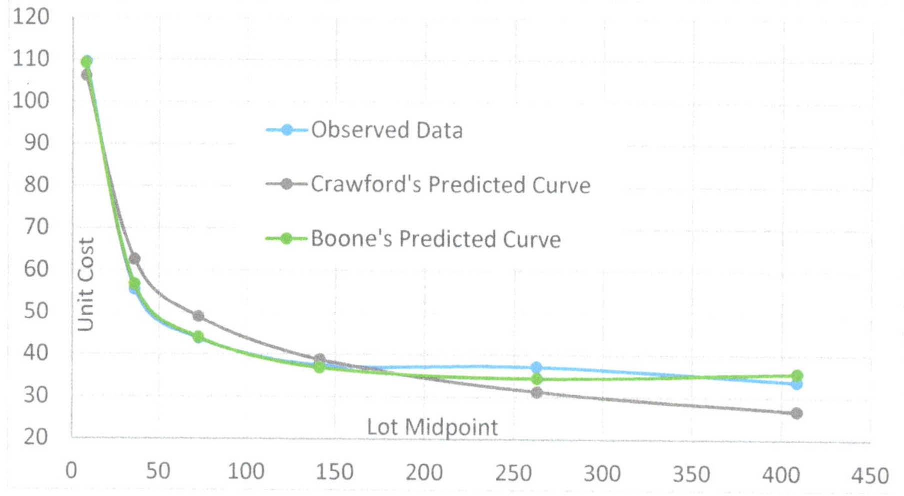

Figure 5. Comparison of Program 1 PME Air Vehicle.

The results of these paired difference tests indicate the improvement with Boone's model is statistically significant. Again, the RMSE percent difference (for labor hours) used a Wilcoxon rank sum test (due to the failure of the Shapiro–Wilk test with a *p*-value less than 0.001).

5. Conclusions

A large, diverse dataset of DoD production programs was used to test if Boone's learning curve more accurately explained error in comparison to traditional learning curve theories. The direct recurring cost data from bomber, cargo, and fighter aircraft along with missiles and munitions programs in units of total dollars and labor hours were analyzed using Cumulative Average and Unit Learning Curve theories. Various components of these programs were analyzed from wings and data link systems to the airframes and air vehicles. Boone's learning curve was tested against both cumulative average and unit learning curve theories using two different measures of model error that resulted in six paired difference tests. This methodology resulted in 998 total observations across all measures and ensured the generalizability of Boone's learning curve was tested.

Boone's learning curve improved upon the traditional learning curve estimates for approximately 42% of the sampled program components while approximately equaling the traditional learning curve error for approximately 51% of program components. Boone's learning curve resulted in a range of mean percentage difference reductions of 6% to 18.6% across all measures. The standard deviations of these improvements were high with coefficients of variation ranging from 150% to 247% across all measures. Absent additional analysis, these high amounts of variability make it challenging to conclude the degree to which Boone's learning curve will improve the accuracy of explaining program component costs in comparison to the traditional estimation methods. Specifically, more research is needed to understand the shape of the learning curve and how it behaves related to production circumstances. It remains unclear which programs are more accurately modelled using Boone's learning curve and to what degree Boone's learning curve will more accurately model program component costs.

The paired difference tests between Boone's learning curve and the traditional theories indicate that Boone's learning curve reduces error to a significant degree across a wide range of measures. Five of the six paired difference tests resulted in rejecting the null hypothesis that Boone's learning curve had an equal amount or more error than the traditional theories at a significance level of 0.05.

Due to data availability, program lot data was used instead of unitary data. Although Boone's learning curve should perform just as well using either type of data, this research cannot conclusively state that Boone's learning curve will more accurately explain programs in unitary data. Also, the majority of data utilized were end-item components in units of total dollars. The total dollar cost includes all cost categories rather than solely labor costs. These data are not ideal when applying learning curve theory and may bias learning curves to display diminishing rates of learning. Despite these potential issues, total dollar cost data are regularly utilized by cost estimators in the field due to data availability. Therefore, the practical applications of this analysis remain valid despite the limitations of using imperfect total dollar cost data in learning curve analysis.

Boone's learning curve was tested on programs whose lot costs were already known and whose parameters can be directly estimated. In other words, Boone's learning curve was tested against the traditional theories on how well it explained rather than predicted program costs. In order to utilize Boone's learning curve to predict costs, a decay value would be selected a priori. Similar to the learning curve slope, an analyst could use the decay value from similar programs to provide a range values to make predictions. Additionally, future research should investigate if Boone's Decay Value can be predicted using various attributes of a program. Tests could be performed on how well Boone's learning curve predicts costs for a program using analogous programs in comparison to the traditional theories. Lastly, additional labor hour data should be collected and analyzed in order to dispel the potential bias of learning curves displaying diminishing rates of learning when analyzed in units of total dollars.

Author Contributions: Conceptualization, D.H., J.E., C.K. and J.R.; methodology, D.H., J.E., C.K., J.R. and A.B.; software, D.H. and S.V.; validation, D.H., J.E., C.K. and J.R.; formal analysis, D.H., J.E., C.K. and J.R.; investigation, D.H., J.E., C.K., J.R. and S.V.; resources, D.H., A.B. and S.V.; data curation, D.H., J.E. and S.V.; writing—original draft preparation, D.H., J.E. and C.K.; writing—review and editing D.H., J.E., C.K., J.R. and A.B.; visualization, D.H. and A.B.; supervision, J.E., C.K., J.R. and A.B.; project administration, D.H., J.E., C.K. and J.R.; funding acquisition, J.E., J.R. and A.B. All authors have read and agreed to the published version of the manuscript.

Funding: This research received no external funding.

Conflicts of Interest: The authors declare no conflict of interest.

Appendix A. Calculation Process for Lot Midpoint Estimation

The following process was implemented to estimate parameters for lot midpoint estimation.

1. Parameter-free lot midpoint approximations (Equation (4)) were calculated for each production lot.
2. Crawford's learning curve parameters A and b were initially estimated using OLS regression.

a. Average unit cost was the dependent variable while lot midpoint, calculated in Step 1, was the independent variable.

3. These initial learning curve parameter estimates were used as starting values to more precisely estimate Crawford's learning curve parameters using GRG non-linear solver. This process generated intermediate estimates of Crawford's learning curve parameters.
4. The intermediate estimate of Crawford's learning curve b parameter was used to calculate a more precise set of lot midpoints using Asher's approximation (Equation (5)).
5. Applying these more precise lot midpoint approximations, Crawford's learning curve parameters A and b were more accurately estimated using GRG nonlinear solver.

Steps 4 and 5 were repeated until the iterative process converged on a solution to produce final estimates of Crawford's learning curve parameters and lot midpoint approximations.

Appendix B. Learning Curve Error Comparisons Using Cumulative Average and Unit Theories

Table A1. Error Comparison using Cumulative Average Theory.

Program	Number of Lots	Number of Units	Component Estimated	Units	Traditional RMSE	Boone RMSE	RMSE Percentage Difference	Traditional MAPE	Boone MAPE	MAPE Percentage Difference
Program 1	6	483	PME-Air Vehicle	Dollars	557.9	111.7	80.0%	3.6%	0.7%	80.9%
Program 1	6	483	PME-Air Vehicle	Hours	15.5	0.3	98.0%	27.2%	0.5%	98.2%
Program 1	6	483	Airframe	Dollars	411.2	114.1	72.3%	2.8%	0.7%	74.7%
Program 1	6	483	Airframe	Hours	21.7	1.5	93.0%	31.0%	1.7%	94.6%
Program 2	5	638	PME-Air Vehicle	Dollars	129.8	6.5	95.0%	2.6%	0.1%	95.6%
Program 3	5	500	PME-Air Vehicle	Dollars	1630.3	291.1	82.1%	20.8%	3.9%	81.5%
Program 4	19	205	PME-Air Vehicle	Dollars	581.7	581.8	0.0%	3.1%	3.1%	0.0%
Program 4	19	205	Airframe	Dollars	546.0	546.4	−0.1%	3.2%	3.2%	−0.1%
Program 5	7	459	PME-Air Vehicle	Dollars	400.8	44.7	88.8%	2.7%	0.3%	88.2%
Program 5	7	459	Electronic Warfare (1)	Dollars	4.8	3.2	32.3%	7.2%	4.8%	33.7%
Program 6	6	98	PME-Air Vehicle	Dollars	99.3	32.2	67.6%	1.1%	0.3%	69.4%
Program 6	6	98	Electronic Warfare (1)	Dollars	12.7	1.7	86.8%	3.6%	0.6%	82.4%
Program 6	6	98	Electronic Warfare (2)	Dollars	15.0	13.3	11.4%	2.3%	2.0%	12.9%
Program 6	6	98	Electronic Warfare (3)	Dollars	1.8	1.1	40.3%	1.3%	0.8%	39.6%
Program 7	7	110	PME-Air Vehicle	Dollars	145.0	98.3	32.2%	1.0%	0.7%	32.6%
Program 7	7	110	Electronic Warfare (1)	Dollars	8.4	3.6	57.2%	2.7%	1.0%	61.3%
Program 7	7	110	Electronic Warfare (2)	Dollars	140.3	107.2	23.6%	1.2%	0.8%	27.5%
Program 7	7	110	Electronic Warfare (3)	Dollars	0.9	0.9	0.0%	0.5%	0.5%	−0.1%
Program 7	7	110	Electronic Warfare (4)	Dollars	140.7	111.3	20.9%	1.3%	1.0%	24.2%
Program 7	7	110	Electronic Warfare (5)	Dollars	21.3	21.0	1.1%	2.2%	2.1%	5.2%
Program 8	8	3529	PME-Air Vehicle	Dollars	27.7	23.6	14.8%	1.4%	1.3%	7.8%
Program 8	8	3529	PME-Air Vehicle	Hours	0.1	0.1	−27.5%	1.1%	1.3%	−27.9%
Program 9	9	3798	PME-Air Vehicle	Dollars	166.5	170.7	−2.5%	8.4%	8.8%	−3.7%
Program 10	10	3803	PME-Air Vehicle	Dollars	8.0	4.8	39.6%	2.5%	1.2%	51.7%
Program 10	10	3803	PME-Air Vehicle	Hours	24.4	14.0	42.7%	4.3%	2.0%	54.0%

Table A1. *Cont.*

Program	Number of Lots	Number of Units	Component Estimated	Units	Traditional RMSE	Boone RMSE	RMSE Percentage Difference	Traditional MAPE	Boone MAPE	MAPE Percentage Difference
Program 11	6	180	PME–Air Vehicle	Dollars	514.0	508.4	1.1%	0.9%	0.8%	4.2%
Program 12	10	20	PME–Air Vehicle	Dollars	699.2	694.1	0.7%	5.8%	5.7%	1.0%
Program 12	10	20	PME–Air Vehicle	Hours	1042.5	906.5	13.1%	9.5%	8.4%	11.8%
Program 12	7	11	Mission Computer (1)	Dollars	44.3	44.3	0.0%	2.5%	2.5%	0.0%
Program 13	5	100	PME–Air Vehicle	Dollars	53,386.7	21,143.7	60.4%	12.8%	4.8%	62.1%
Program 13	5	100	Airframe	Dollars	6569.7	6578.0	−0.1%	3.7%	3.7%	0.0%
Program 14	5	275	PME–Air Vehicle	Dollars	3114.0	145.5	95.3%	3.8%	0.2%	95.5%
Program 15	10	77	PME–Air Vehicle	Dollars	44,386.0	44,390.2	0.0%	9.5%	9.5%	0.0%
Program 15	12	83	PME–Air Vehicle	Hours	79,242.0	79,247.5	0.0%	6.5%	6.5%	0.0%
Program 15	11	83	Airframe	Dollars	39,624.4	39,628.0	0.0%	10.6%	10.6%	0.0%
Program 15	10	68	Mission Computer (1)	Dollars	1959.3	1959.4	0.0%	17.0%	17.0%	0.0%
Program 16	9	76	PME–Air Vehicle	Dollars	436.3	144.4	66.9%	2.6%	1.0%	62.9%
Program 17	5	50	PME–Air Vehicle	Dollars	13,023.6	13,029.8	0.0%	2.8%	2.8%	−0.1%
Program 18	9	31	PME–Air Vehicle	Dollars	2942.5	2941.9	0.0%	1.0%	0.9%	0.0%
Program 19	6	98	PME–Air Vehicle	Dollars	313.3	313.4	0.0%	0.5%	0.5%	−0.1%
Program 20	11	84	PME–Air Vehicle	Dollars	1568.7	1121.9	28.5%	1.7%	1.5%	7.8%
Program 20	7	59	Electronic Warfare (1)	Dollars	452.8	143.0	68.4%	4.6%	1.3%	71.5%
Program 20	11	84	Electronic Warfare (2)	Dollars	98.7	76.5	22.5%	3.4%	3.6%	−6.3%
Program 20	7	59	Electronic Warfare (5)	Dollars	562.5	517.4	8.0%	1.8%	1.8%	1.7%
Program 21	6	326	PME–Air Vehicle	Dollars	5267.1	2408.8	54.3%	8.0%	4.2%	47.4%
Program 21	7	344	Airframe	Dollars	4819.5	2544.3	47.2%	9.1%	5.4%	40.4%
Program 21	7	344	Avionics	Dollars	763.2	429.9	43.7%	6.6%	3.9%	40.8%
Program 21	14	453	PME–Air Vehicle	Hours	3493.6	3495.9	−0.1%	4.8%	4.8%	0.1%
Program 21	14	453	Airframe	Hours	4338.4	4339.7	0.0%	6.2%	6.2%	0.1%
Program 22	8	538	PME–Air Vehicle	Hours	856.7	857.7	−0.1%	2.5%	2.6%	−0.1%
Program 22	8	538	Airframe	Hours	5608.5	5609.7	0.0%	15.8%	15.9%	−0.1%
Program 23	5	469	PME–Air Vehicle	Dollars	637.5	339.3	46.8%	5.4%	2.9%	47.3%
Program 24	10	59	PME–Air Vehicle	Dollars	3032.5	3033.0	0.0%	2.2%	2.2%	0.0%
Program 25	9	348	PME–Air Vehicle	Dollars	117.8	118.1	−0.2%	0.9%	0.9%	−0.2%
Program 26	5	109	PME–Air Vehicle	Dollars	3247.4	1676.8	48.4%	11.0%	6.0%	45.7%
Program 26	5	109	PME–Air Vehicle	Hours	607.1	453.5	25.3%	5.7%	4.2%	25.9%
Program 27	18	631	PME–Air Vehicle	Dollars	1669.6	913.3	45.3%	3.6%	1.9%	46.2%
Program 28	6	425	PME–Air Vehicle	Dollars	320.0	322.0	−0.6%	0.9%	0.9%	−0.6%
Program 28	7	522	PME–Air Vehicle	Hours	1776.1	1785.6	−0.5%	1.8%	1.8%	−0.1%
Program 28	7	522	Airframe	Hours	1389.9	1393.9	−0.3%	1.2%	1.2%	−0.2%
Program 29	9	358	PME–Air Vehicle	Hours	610.6	611.1	−0.1%	0.9%	0.9%	0.4%
Program 29	9	358	Airframe	Hours	4804.8	2124.2	55.8%	7.3%	2.9%	60.1%
Program 30	5	204	PME–Air Vehicle	Dollars	513.5	212.7	58.6%	1.2%	0.5%	56.1%
Program 31	5	605	PME–Air Vehicle	Dollars	1482.6	629.1	57.6%	6.1%	2.9%	53.1%

Table A1. *Cont.*

Program	Number of Lots	Number of Units	Component Estimated	Units	Traditional RMSE	Boone RMSE	RMSE Percentage Difference	Traditional MAPE	Boone MAPE	MAPE Percentage Difference
Program 32	5	870	PME–Air Vehicle	Dollars	61.3	61.6	−0.5%	0.4%	0.4%	−0.3%
Program 33	10	178	PME–Air Vehicle	Dollars	7093.5	7101.1	−0.1%	3.5%	3.5%	−0.1%
Program 33	10	178	PME–Air Vehicle	Hours	8131.1	8144.1	−0.2%	2.9%	2.9%	−0.1%
Program 33	10	178	Airframe	Dollars	1906.9	1910.8	−0.2%	1.7%	1.7%	−0.2%
Program 33	10	712	Body	Dollars	232.2	234.9	−1.2%	1.5%	1.6%	−1.3%
Program 33	10	178	Alighting Gear	Dollars	76.6	76.6	0.0%	7.9%	7.9%	0.0%
Program 33	10	178	Auxiliary Power Plant	Dollars	90.7	90.7	−0.1%	3.9%	3.9%	−0.1%
Program 33	10	178	Electronic Warfare (1)	Dollars	775.5	776.1	−0.1%	6.5%	6.5%	−0.1%
Program 33	10	178	Electronic Warfare (2)	Dollars	360.1	273.4	24.1%	58.3%	46.0%	21.2%
Program 33	10	178	Electronic Warfare (3)	Dollars	62.5	62.4	0.2%	5.7%	5.7%	0.1%
Program 33	10	178	Empennage	Dollars	352.2	352.3	0.0%	5.1%	5.1%	−0.1%
Program 33	10	178	Hydraulic	Dollars	22.7	22.7	−0.1%	2.2%	2.2%	−0.1%
Program 33	10	178	Wing	Dollars	296.5	296.9	−0.1%	2.3%	2.3%	−0.1%
Program 34	6	67	PME–Air Vehicle	Dollars	11,059.1	11,061.2	0.0%	6.6%	6.6%	0.0%
Program 34	6	67	PME–Air Vehicle	Hours	9058.6	9061.7	0.0%	4.4%	4.4%	0.0%
Program 34	6	67	Airframe	Dollars	2798.1	2004.6	28.4%	2.8%	1.7%	37.9%
Program 34	6	201	Body	Dollars	1924.5	828.9	56.9%	19.0%	8.7%	54.0%
Program 34	6	67	Alighting Gear	Dollars	316.5	166.9	47.3%	17.2%	8.3%	51.9%
Program 34	6	67	Electrical	Dollars	50.7	50.7	−0.1%	1.9%	1.9%	−0.1%
Program 34	6	67	Electronic Warfare (1)	Dollars	428.3	428.4	0.0%	5.3%	5.3%	0.0%
Program 34	5	49	Empennage	Dollars	202.2	202.2	0.0%	4.1%	4.1%	0.0%
Program 34	6	67	EO/IR	Dollars	45.6	36.6	19.7%	1.2%	1.1%	13.1%
Program 34	6	67	EOTS	Dollars	347.6	347.7	0.0%	6.5%	6.5%	0.0%
Program 34	6	67	Hydraulic	Dollars	122.3	101.5	17.0%	8.4%	6.2%	26.8%
Program 34	6	67	Mission Computer (1)	Dollars	484.8	484.9	0.0%	0.9%	0.9%	−0.2%
Program 34	6	67	Surface Controls	Dollars	196.0	196.0	0.0%	4.9%	4.9%	0.0%
Program 34	6	67	Wing	Dollars	998.4	998.6	0.0%	3.3%	3.3%	−0.1%
Program 35	5	41	PME–Air Vehicle	Dollars	3578.6	3579.8	0.0%	1.5%	1.5%	0.0%
Program 35	5	41	PME–Air Vehicle	Hours	2003.7	2004.7	0.0%	1.1%	1.1%	0.0%
Program 35	5	50	Airframe	Dollars	609.3	610.4	−0.2%	0.6%	0.6%	−0.3%
Program 35	5	150	Body	Dollars	235.8	156.5	33.6%	1.9%	1.4%	28.0%
Program 35	5	50	Alighting Gear	Dollars	13.2	13.2	−0.1%	0.5%	0.5%	0.0%
Program 35	5	50	Electronic Warfare (1)	Dollars	259.6	259.7	0.0%	3.2%	3.2%	0.0%
Program 35	5	50	EO/IR	Dollars	121.6	121.7	0.0%	1.3%	1.3%	−0.1%
Program 35	5	50	EOTS	Dollars	177.9	177.9	0.0%	2.8%	2.8%	−0.1%
Program 35	5	50	Hydraulic	Dollars	58.2	58.2	0.0%	3.1%	3.1%	0.0%
Program 35	5	50	Radar	Dollars	256.8	256.9	0.0%	3.2%	3.2%	0.0%
Program 35	5	50	Surface Controls	Dollars	121.5	121.5	0.0%	2.6%	2.6%	0.0%
Program 35	5	50	Wing	Dollars	1213.5	1213.6	0.0%	3.8%	3.8%	0.0%
Program 36	13	1285	PME–Air Vehicle	Dollars	28.8	29.4	−2.1%	0.6%	0.6%	−2.2%
Program 37	6	432	PME–Air Vehicle	Dollars	791.3	793.8	−0.3%	3.4%	3.4%	−0.4%
Program 38	6	52	PME–Air Vehicle	Dollars	253.6	154.9	38.9%	1.2%	0.7%	41.6%
Program 38	6	44	PME–Air Vehicle	Hours	831.5	614.2	26.1%	1.3%	0.8%	42.8%
Program 39	19	1023	PME–Air Vehicle	Dollars	19.3	19.3	−0.2%	0.7%	0.7%	−0.2%
Program 40	5	1725	PME–Air Vehicle	Dollars	19.2	0.6	96.7%	2.0%	0.1%	97.0%
Program 41	10	16	PME–Air Vehicle	Dollars	14,787.6	14,787.8	0.0%	5.2%	5.2%	0.0%

Table A1. *Cont.*

Program	Number of Lots	Number of Units	Component Estimated	Units	Traditional RMSE	Boone RMSE	RMSE Percentage Difference	Traditional MAPE	Boone MAPE	MAPE Percentage Difference
Program 41	10	16	Data Link (1)	Dollars	138.8	138.8	0.0%	3.7%	3.7%	0.0%
Program 42	11	203	PME–Air Vehicle	Dollars	1000.0	1000.1	0.0%	7.0%	7.0%	0.0%
Program 42	11	899	Electronic Warfare (1)	Dollars	67.5	67.7	−0.2%	13.9%	13.9%	−0.5%
Program 43	11	203	PME–Air Vehicle	Dollars	1121.7	1121.9	0.0%	5.5%	5.5%	0.0%
Program 43	13	251	PME–Air Vehicle	Hours	1944.2	1762.2	9.4%	3.4%	3.2%	6.1%
Program 44	5	136	PME–Air Vehicle	Dollars	57.1	16.3	71.4%	1.1%	0.3%	71.4%
Program 45	9	155	PME–Air Vehicle	Dollars	149.6	149.7	−0.1%	0.3%	0.3%	−0.1%
Program 46	6	68	PME–Air Vehicle	Dollars	3435.9	3436.0	0.0%	1.7%	1.7%	0.1%
Program 46	6	68	PME–Air Vehicle	Hours	2286.4	2286.6	0.0%	2.6%	2.6%	0.0%
Program 46	6	68	Airframe	Dollars	539.1	527.6	2.1%	2.3%	2.1%	10.9%
Program 46	6	68	Data Link (1)	Dollars	44.0	44.0	0.0%	3.0%	3.0%	0.0%
Program 46	6	68	Electronic Warfare (1)	Dollars	221.8	221.9	0.0%	5.4%	5.4%	0.0%
Program 46	6	68	Electronic Warfare (2)	Dollars	220.0	220.0	0.0%	6.5%	6.5%	0.0%
Program 46	6	68	Electronic Warfare (3)	Dollars	17.7	8.8	50.4%	2.2%	1.0%	54.6%
Program 46	6	68	Electronic Warfare (4)	Dollars	530.0	530.0	0.0%	5.2%	5.2%	0.0%
Program 46	6	68	EO/IR	Dollars	120.7	120.8	0.0%	15.7%	15.7%	0.0%
Program 46	6	68	Mission Computer (1)	Dollars	477.9	478.0	0.0%	4.3%	4.3%	0.0%
Program 47	9	36	PME–Air Vehicle	Dollars	1039.4	1039.4	0.0%	2.5%	2.5%	0.0%
Program 47	9	36	PME–Air Vehicle	Hours	8278.7	8278.6	0.0%	15.5%	15.5%	0.0%
Program 47	9	36	Data Link (1)	Dollars	170.2	170.2	0.0%	17.7%	17.7%	0.0%
Program 48	5	179	PME–Air Vehicle	Dollars	1858.3	391.3	78.9%	3.1%	0.6%	79.4%
Program 49	6	180	PME–Air Vehicle	Dollars	435.3	99.8	77.1%	4.4%	1.0%	76.5%
Program 50	5	488	PME–Air Vehicle	Dollars	349.3	350.7	−0.4%	3.3%	3.4%	−0.8%
Program 51	6	663	PME–Air Vehicle	Dollars	5.6	3.6	36.6%	0.6%	0.4%	24.8%
Program 52	5	380	PME–Air Vehicle	Dollars	456.9	454.6	0.5%	9.0%	8.9%	0.3%
Program 53	6	749	PME–Air Vehicle	Dollars	37.2	36.6	1.7%	0.5%	0.5%	4.3%
Program 54	8	194	PME–Air Vehicle	Dollars	28.8	28.8	−0.1%	0.6%	0.6%	−0.1%
Program 55	9	677	PME–Air Vehicle	Dollars	74.8	74.8	0.0%	1.6%	1.6%	0.0%
Program 56	5	590	PME–Air Vehicle	Dollars	6.6	6.6	0.5%	0.2%	0.2%	6.3%
Program 57	5	579	PME–Air Vehicle	Dollars	22.8	22.8	−0.1%	0.8%	0.8%	0.0%

Table A2. Error Comparison using Unit Theory.

Program	Number of Lots	Number of Units	Component Estimated	Units	Traditional RMSE	Boone RMSE	RMSE Percentage Difference	Traditional MAPE	Boone MAPE	MAPE Percentage Difference
Program 1	7	503	Airframe	Hours	4.6	3.5	23.4%	7.1%	5.0%	28.7%
Program 1	6	483	PME–Air Vehicle	Hours	5.4	1.5	72.5%	11.3%	2.9%	74.0%
Program 1	7	503	PME–Air Vehicle	Dollars	2260.6	517.0	77.1%	12.9%	3.2%	75.2%
Program 1	7	503	Airframe	Dollars	2383.2	857.9	64.0%	14.6%	4.9%	66.4%

Table A2. *Cont.*

Program	Number of Lots	Number of Units	Component Estimated	Units	Traditional RMSE	Boone RMSE	RMSE Percentage Difference	Traditional MAPE	Boone MAPE	MAPE Percentage Difference
Program 2	5	638	PME–Air Vehicle	Dollars	315.4	195.3	38.1%	5.8%	4.3%	26.3%
Program 3	5	500	PME–Air Vehicle	Dollars	2984.5	1120.2	62.5%	49.4%	17.6%	64.4%
Program 4	7	357	Airframe	Dollars	2662.2	2664.3	−0.1%	13.1%	13.2%	−0.1%
Program 4	9	424	PME–Air Vehicle	Dollars	9323.3	4999.8	46.4%	37.9%	14.1%	62.8%
Program 5	19	205	Airframe	Dollars	2446.1	2445.8	0.0%	12.6%	12.6%	−0.3%
Program 5	19	205	PME–Air Vehicle	Dollars	3228.6	3228.9	0.0%	12.4%	12.4%	0.0%
Program 6	7	459	Electronic Warfare (1)	Dollars	20.9	20.9	0.0%	30.8%	30.8%	0.0%
Program 6	7	459	PME–Air Vehicle	Dollars	1439.9	738.1	48.7%	11.3%	5.9%	47.2%
Program 7	5	321	PME–Air Vehicle	Dollars	37.9	33.3	12.2%	3.8%	3.8%	1.1%
Program 8	6	98	Electronic Warfare (3)	Dollars	5.2	4.9	6.1%	4.8%	4.8%	1.4%
Program 8	6	98	Electronic Warfare (2)	Dollars	84.2	70.3	16.5%	11.1%	10.6%	4.7%
Program 8	6	98	PME–Air Vehicle	Dollars	375.2	339.5	9.5%	4.2%	3.7%	13.4%
Program 8	6	98	Electronic Warfare (1)	Dollars	27.5	18.7	31.9%	10.2%	5.9%	42.5%
Program 9	7	110	Electronic Warfare (5)	Dollars	102.9	99.2	3.5%	9.7%	10.4%	−6.6%
Program 9	7	110	Electronic Warfare (3)	Dollars	6.4	6.4	0.0%	4.7%	4.7%	0.0%
Program 9	7	110	Electronic Warfare (4)	Dollars	653.6	653.6	0.0%	6.2%	6.2%	0.0%
Program 9	7	110	Electronic Warfare (2)	Dollars	709.4	709.4	0.0%	6.1%	6.1%	0.0%
Program 9	7	110	PME–Air Vehicle	Dollars	668.5	668.5	0.0%	5.1%	5.1%	0.0%
Program 9	7	110	Electronic Warfare (1)	Dollars	31.6	29.1	8.0%	8.7%	8.0%	8.3%
Program 10	9	1586	PME–Air Vehicle	Dollars	115.5	115.6	−0.2%	12.5%	12.5%	−0.2%
Program 10	10	1796	PME–Air Vehicle	Hours	150.8	150.9	0.0%	12.5%	12.5%	−0.1%
Program 11	8	3529	PME–Air Vehicle	Hours	0.9	0.7	21.2%	27.5%	44.9%	−63.4%
Program 11	8	3529	PME–Air Vehicle	Dollars	97.1	97.5	−0.4%	10.1%	10.4%	−2.1%
Program 12	16	7891	PME–Air Vehicle	Hours	520.1	525.6	−1.1%	86.2%	86.2%	0.0%
Program 12	21	10035	PME–Air Vehicle	Dollars	243.8	239.2	1.9%	30.1%	28.8%	4.2%
Program 13	6	3385	EO	Dollars	12.1	9.4	22.5%	10.7%	9.6%	10.0%
Program 13	10	3803	PME–Air Vehicle	Dollars	33.6	24.8	26.1%	10.3%	7.5%	27.1%
Program 13	10	3803	PME–Air Vehicle	Hours	130.1	100.5	22.7%	21.5%	17.1%	20.7%
Program 14	6	180	PME–Air Vehicle	Dollars	2249.4	1008.9	55.2%	6.4%	2.3%	64.2%
Program 15	10	20	PME–Air Vehicle	Hours	3430.3	3430.4	0.0%	41.5%	41.5%	0.0%
Program 15	10	20	PME–Air Vehicle	Dollars	3013.9	3013.9	0.0%	17.4%	17.4%	0.0%
Program 15	7	11	Mission Computer (1)	Dollars	213.9	213.9	0.0%	11.6%	11.5%	0.6%
Program 16	5	100	Airframe	Dollars	10,807.3	7455.4	31.0%	7.0%	4.1%	41.8%
Program 16	5	100	PME–Air Vehicle	Dollars	137,225.9	81,884.9	40.3%	51.7%	26.9%	48.0%
Program 17	5	275	PME–Air Vehicle	Dollars	8837.5	1396.3	84.2%	17.6%	3.3%	81.6%
Program 18	12	83	PME–Air Vehicle	Hours	266,012.8	266,015.3	0.0%	39.3%	39.3%	0.0%
Program 18	11	83	Airframe	Dollars	89,956.0	89,961.1	0.0%	39.1%	39.1%	0.0%
Program 18	10	68	Mission Computer (1)	Dollars	4143.0	4143.2	0.0%	68.2%	68.2%	0.0%

Table A2. *Cont.*

Program	Number of Lots	Number of Units	Component Estimated	Units	Traditional RMSE	Boone RMSE	RMSE Percentage Difference	Traditional MAPE	Boone MAPE	MAPE Percentage Difference
Program 18	11	83	PME–Air Vehicle	Dollars	82,138.6	82,143.3	0.0%	23.2%	23.2%	0.0%
Program 19	5	45	Airframe	Dollars	501.2	501.2	0.0%	53.9%	53.9%	0.0%
Program 19	5	45	PME–Air Vehicle	Dollars	649.0	649.0	0.0%	17.6%	17.6%	0.0%
Program 19	5	45	Mission Computer (1)	Dollars	61.7	59.7	3.2%	9.8%	9.7%	1.2%
Program 20	9	76	PME–Air Vehicle	Dollars	1108.7	522.5	52.9%	7.2%	3.6%	49.9%
Program 21	5	50	PME–Air Vehicle	Dollars	24,625.3	6362.0	74.2%	7.4%	2.3%	69.5%
Program 22	9	31	PME–Air Vehicle	Dollars	16,636.3	16,636.4	0.0%	6.6%	6.6%	0.0%
Program 23	5	14	PME–Air Vehicle	Dollars	14,475.8	14,476.0	0.0%	8.7%	8.7%	0.0%
Program 24	6	98	PME–Air Vehicle	Dollars	2259.9	2260.1	0.0%	3.3%	3.3%	0.0%
Program 25	7	59	Electronic Warfare (5)	Dollars	2808.4	2805.2	0.1%	14.8%	15.4%	−4.0%
Program 25	11	84	PME–Air Vehicle	Dollars	5083.2	4228.8	16.8%	8.7%	9.2%	−5.2%
Program 25	11	84	Electronic Warfare (2)	Dollars	248.9	248.6	0.1%	13.9%	14.3%	−2.9%
Program 25	7	59	Electronic Warfare (1)	Dollars	1259.1	653.3	48.1%	16.1%	7.1%	55.6%
Program 26	7	344	Airframe	Dollars	11,474.7	8294.9	27.7%	22.7%	21.5%	5.3%
Program 26	7	344	Avionics	Dollars	2218.8	2102.8	5.2%	29.5%	26.9%	8.8%
Program 26	7	344	PME–Air Vehicle	Dollars	12,898.4	8742.1	32.2%	20.7%	16.9%	18.4%
Program 27	14	453	PME–Air Vehicle	Hours	54,142.9	53,766.4	0.7%	59.9%	63.1%	−5.4%
Program 27	14	453	Airframe	Hours	70,415.0	69,426.8	1.4%	58.8%	59.1%	−0.5%
Program 28	8	538	PME–Air Vehicle	Hours	3828.8	3829.8	0.0%	9.8%	9.9%	0.0%
Program 28	8	538	Airframe	Hours	3865.3	3866.2	0.0%	7.6%	7.6%	0.0%
Program 29	8	529	Hydraulic	Dollars	156.9	156.4	0.3%	22.3%	22.9%	−2.8%
Program 29	12	477	Airframe	Dollars	6490.2	5974.2	7.9%	14.2%	14.4%	−1.8%
Program 29	12	477	Wing	Dollars	712.3	712.7	−0.1%	27.8%	27.8%	−0.1%
Program 29	11	433	Electronic Warfare (1)	Dollars	57.5	57.5	0.0%	13.5%	13.5%	−0.1%
Program 29	8	309	Electrical	Dollars	230.6	230.7	−0.1%	8.2%	8.2%	0.0%
Program 29	12	1045	Body	Dollars	1922.2	1826.7	5.0%	26.0%	25.9%	0.7%
Program 29	5	177	Empennage	Dollars	32.3	22.0	31.8%	6.1%	4.6%	24.5%
Program 29	12	477	PME–Air Vehicle	Dollars	8218.5	5525.3	32.8%	15.0%	10.2%	32.0%
Program 29	8	309	Alighting Gear	Dollars	205.7	42.2	79.5%	11.6%	2.0%	83.1%
Program 30	5	469	PME–Air Vehicle	Dollars	1283.8	891.8	30.5%	13.5%	8.3%	38.3%
Program 31	10	59	PME–Air Vehicle	Dollars	11,978.9	11,979.3	0.0%	8.6%	8.6%	0.0%
Program 32	9	348	PME–Air Vehicle	Dollars	430.6	430.8	0.0%	3.5%	3.5%	−0.1%
Program 33	5	109	PME–Air Vehicle	Hours	993.9	994.0	0.0%	9.5%	9.5%	0.0%
Program 33	5	109	PME–Air Vehicle	Dollars	6824.7	6824.8	0.0%	28.2%	28.2%	0.0%
Program 34	18	631	PME–Air Vehicle	Dollars	6926.7	2799.9	59.6%	17.0%	6.6%	61.0%
Program 35	6	425	PME–Air Vehicle	Dollars	1135.8	1137.5	−0.2%	3.5%	3.5%	−0.2%
Program 35	7	522	PME–Air Vehicle	Hours	4615.3	4458.5	3.4%	6.3%	6.1%	3.1%
Program 35	7	522	Airframe	Hours	6757.0	6280.7	7.0%	5.7%	5.4%	4.8%
Program 36	9	358	PME–Air Vehicle	Hours	5118.7	5120.1	0.0%	6.8%	6.8%	0.0%
Program 36	9	358	Airframe	Hours	12,155.2	11,257.1	7.4%	15.5%	14.3%	7.6%
Program 37	5	204	PME–Air Vehicle	Dollars	1468.7	921.0	37.3%	2.9%	1.9%	36.4%
Program 38	5	605	PME–Air Vehicle	Dollars	2641.9	1527.7	42.2%	14.9%	8.1%	46.0%

Table A2. *Cont.*

Program	Number of Lots	Number of Units	Component Estimated	Units	Traditional RMSE	Boone RMSE	RMSE Percentage Difference	Traditional MAPE	Boone MAPE	MAPE Percentage Difference
Program 39	5	870	PME–Air Vehicle	Dollars	310.9	311.5	−0.2%	2.3%	2.3%	−0.2%
Program 40	10	178	Electronic Warfare (3)	Dollars	751.2	551.9	26.5%	69.7%	74.7%	−7.1%
Program 40	10	712	Body	Dollars	617.6	577.6	6.5%	4.8%	5.1%	−7.6%
Program 40	10	178	Airframe	Dollars	4251.9	4226.4	0.6%	4.8%	4.9%	−1.0%
Program 40	10	178	Electronic Warfare (2)	Dollars	721.7	721.7	0.0%	393.4%	393.4%	0.0%
Program 40	10	178	Electronic Warfare (1)	Dollars	1642.3	1643.0	0.0%	20.7%	20.7%	0.0%
Program 40	10	178	PME–Air Vehicle	Hours	13,454.5	13,466.8	−0.1%	6.0%	6.0%	−0.1%
Program 40	10	178	Auxiliary Power Plant	Dollars	385.1	385.1	0.0%	24.9%	24.9%	0.0%
Program 40	10	178	PME–Air Vehicle	Dollars	12,231.7	12,236.6	0.0%	7.9%	7.9%	0.0%
Program 40	10	178	Alighting Gear	Dollars	233.6	233.6	0.0%	30.1%	30.1%	0.0%
Program 40	10	178	Wing	Dollars	607.4	607.6	0.0%	6.2%	6.2%	0.0%
Program 40	10	178	Empennage	Dollars	702.1	702.1	0.0%	17.4%	17.4%	0.0%
Program 40	10	178	Hydraulic	Dollars	72.2	70.2	2.8%	9.0%	8.8%	2.2%
Program 41	6	67	PME–Air Vehicle	Hours	12,741.5	12,743.8	0.0%	9.5%	9.5%	0.0%
Program 41	5	49	Empennage	Dollars	242.2	242.2	0.0%	5.8%	5.9%	0.0%
Program 41	6	67	PME–Air Vehicle	Dollars	16,643.9	16,645.6	0.0%	10.7%	10.7%	0.0%
Program 41	6	67	Surface Controls	Dollars	281.7	281.7	0.0%	7.7%	7.7%	0.0%
Program 41	6	67	EOTS	Dollars	442.3	442.4	0.0%	9.5%	9.5%	0.0%
Program 41	6	67	Wing	Dollars	1927.0	1927.3	0.0%	7.4%	7.4%	0.0%
Program 41	6	67	Electrical	Dollars	57.2	57.2	0.0%	2.1%	2.1%	0.0%
Program 41	6	67	Electronic Warfare (1)	Dollars	547.3	547.3	0.0%	8.1%	8.1%	0.0%
Program 41	6	67	Hydraulic	Dollars	281.5	274.6	2.4%	19.4%	19.0%	2.0%
Program 41	6	67	Mission Computer (1)	Dollars	1698.1	1542.4	9.2%	4.6%	3.7%	19.5%
Program 41	6	67	Airframe	Dollars	6877.8	5547.4	19.3%	8.7%	6.4%	26.8%
Program 41	6	67	Alighting Gear	Dollars	582.3	521.1	10.5%	28.3%	25.0%	11.6%
Program 41	6	67	EO/IR	Dollars	233.0	89.4	61.6%	9.3%	3.1%	66.8%
Program 41	6	201	Body	Dollars	3431.8	2343.2	31.7%	42.6%	29.9%	29.7%
Program 42	5	41	PME–Air Vehicle	Dollars	8498.6	8499.6	0.0%	6.2%	6.2%	0.0%
Program 42	5	41	PME–Air Vehicle	Hours	15,696.5	15,696.9	0.0%	10.7%	10.7%	0.0%
Program 42	5	50	EOTS	Dollars	593.3	593.3	0.0%	11.6%	11.6%	0.0%
Program 42	5	50	EO/IR	Dollars	578.4	578.4	0.0%	7.5%	7.5%	0.0%
Program 42	5	50	Hydraulic	Dollars	297.0	297.0	0.0%	15.4%	15.4%	0.0%
Program 42	5	50	Surface Controls	Dollars	424.9	424.9	0.0%	11.0%	11.0%	0.0%
Program 42	5	50	Radar	Dollars	733.8	733.8	0.0%	10.9%	10.9%	0.0%
Program 42	5	50	Airframe	Dollars	5222.7	5222.8	0.0%	5.9%	5.9%	0.0%
Program 42	5	50	Electronic Warfare (1)	Dollars	746.5	746.5	0.0%	10.7%	10.7%	0.0%
Program 42	5	50	Wing	Dollars	3726.6	3726.7	0.0%	16.5%	16.5%	0.0%
Program 42	5	50	Alighting Gear	Dollars	78.6	77.4	1.5%	3.6%	3.5%	2.3%
Program 42	5	150	Body	Dollars	1588.5	892.1	43.8%	12.6%	8.7%	30.8%
Program 43	13	1285	PME–Air Vehicle	Dollars	88.1	88.8	−0.8%	1.9%	1.9%	−1.0%
Program 44	6	432	PME–Air Vehicle	Dollars	1621.0	1623.3	−0.1%	10.0%	10.0%	−0.2%
Program 45	9	63	PME–Air Vehicle	Dollars	2152.3	1557.1	27.7%	9.5%	6.4%	33.2%
Program 46	6	44	PME–Air Vehicle	Hours	7736.9	7255.3	6.2%	17.6%	16.7%	4.8%
Program 46	10	113	PME–Air Vehicle	Dollars	797.9	627.0	21.4%	3.8%	2.9%	22.7%
Program 47	19	1023	PME–Air Vehicle	Dollars	115.2	115.2	0.0%	4.3%	4.2%	0.2%
Program 48	5	1725	PME–Air Vehicle	Dollars	59.8	3.1	94.9%	6.8%	0.3%	95.4%

Table A2. *Cont.*

Program	Number of Lots	Number of Units	Component Estimated	Units	Traditional RMSE	Boone RMSE	RMSE Percentage Difference	Traditional MAPE	Boone MAPE	MAPE Percentage Difference
Program 49	10	16	Data Link (1)	Dollars	470.3	470.3	0.0%	20.4%	20.4%	0.0%
Program 49	10	16	PME–Air Vehicle	Dollars	41,008.9	41,009.2	0.0%	14.1%	14.1%	0.0%
Program 50	7	577	PME–Air Vehicle	Dollars	1674.7	1224.7	26.9%	5.5%	4.6%	15.7%
Program 51	12	244	PME–Air Vehicle	Hours	625.6	612.8	2.0%	191.4%	191.8%	−0.2%
Program 52	11	899	Electronic Warfare (1)	Dollars	90.1	90.2	−0.1%	29.2%	29.3%	−0.1%
Program 52	11	203	PME–Air Vehicle	Dollars	2995.1	2992.0	0.1%	24.9%	23.6%	5.2%
Program 53	13	251	PME–Air Vehicle	Hours	4585.2	4585.2	0.0%	6.7%	6.7%	0.0%
Program 53	11	203	PME–Air Vehicle	Dollars	2459.9	2460.0	0.0%	9.6%	9.6%	0.0%
Program 54	11	184	PME–Air Vehicle	Hours	7010.4	7010.7	0.0%	18.0%	18.0%	0.0%
Program 54	9	134	PME–Air Vehicle	Dollars	1907.3	970.0	49.1%	11.8%	6.5%	44.9%
Program 55	5	136	PME–Air Vehicle	Dollars	321.6	277.7	13.7%	5.5%	4.7%	14.8%
Program 56	9	155	PME–Air Vehicle	Dollars	1356.5	1356.6	0.0%	3.9%	3.9%	0.0%
Program 57	6	68	EO/IR	Dollars	326.0	326.0	0.0%	1261.8%	1261.8%	0.0%
Program 57	6	68	PME–Air Vehicle	Dollars	8574.7	8470.9	1.2%	4.3%	4.3%	−0.5%
Program 57	6	68	Electronic Warfare (1)	Dollars	998.8	998.9	0.0%	58.9%	58.9%	0.0%
Program 57	6	68	Electronic Warfare (2)	Dollars	750.2	750.2	0.0%	31.3%	31.3%	0.0%
Program 57	6	68	Data Link (1)	Dollars	94.8	94.8	0.0%	7.2%	7.2%	0.0%
Program 57	6	68	Electronic Warfare (4)	Dollars	1156.3	1156.3	0.0%	12.2%	12.2%	0.0%
Program 57	6	68	Mission Computer (1)	Dollars	1030.6	1030.6	0.0%	13.0%	13.0%	0.0%
Program 57	6	68	PME–Air Vehicle	Hours	6435.9	6435.0	0.0%	12.3%	12.3%	0.3%
Program 57	6	68	Airframe	Dollars	1443.2	1285.1	11.0%	6.7%	5.4%	18.5%
Program 57	6	68	Electronic Warfare (3)	Dollars	53.4	21.8	59.1%	7.2%	3.0%	58.5%
Program 58	9	36	PME–Air Vehicle	Hours	60,347.2	60,347.3	0.0%	78.2%	78.2%	0.0%
Program 58	9	36	Data Link (1)	Dollars	227.8	227.8	0.0%	29.3%	29.3%	0.0%
Program 58	9	36	PME–Air Vehicle	Dollars	4570.2	4570.2	0.0%	10.9%	10.9%	0.0%
Program 58	5	18	EO/IR	Dollars	3488.4	3469.8	0.5%	28.8%	28.7%	0.3%
Program 59	5	179	PME–Air Vehicle	Dollars	4583.3	1334.5	70.9%	8.1%	2.8%	65.4%
Program 60	6	180	PME–Air Vehicle	Dollars	1010.5	333.9	67.0%	12.4%	4.6%	63.1%
Program 61	5	488	PME–Air Vehicle	Dollars	502.3	486.5	3.1%	9.2%	7.7%	16.3%
Program 62	6	78	PME–Air Vehicle	Hours	6027.1	5952.3	1.2%	33.8%	34.3%	−1.6%
Program 62	6	97	Airframe	Hours	2648.5	2649.0	0.0%	20.5%	20.5%	0.0%
Program 62	9	110	PME–Air Vehicle	Dollars	13,027.5	13,028.9	0.0%	24.0%	24.0%	0.0%
Program 63	6	663	PME–Air Vehicle	Dollars	23.2	21.1	9.2%	2.9%	2.6%	11.6%
Program 64	5	380	PME–Air Vehicle	Dollars	1520.9	1521.2	0.0%	57.4%	57.4%	0.0%
Program 65	6	749	PME–Air Vehicle	Dollars	116.6	115.9	0.6%	1.7%	1.8%	−5.1%
Program 66	8	194	PME–Air Vehicle	Dollars	128.3	119.3	7.0%	2.6%	2.4%	8.6%
Program 67	9	677	PME–Air Vehicle	Dollars	273.5	273.5	0.0%	5.1%	5.1%	0.0%
Program 68	5	590	PME–Air Vehicle	Dollars	87.1	87.2	0.0%	2.8%	2.8%	0.0%
Program 69	5	579	PME–Air Vehicle	Dollars	305.7	305.8	0.0%	9.5%	9.5%	0.0%

References

1. United States Government Accountability Office; Oakley, S.S. *Weapon Systems Annual Assessment: Knowledge Gaps Pose Risks to Sustaining Recent Positive Trends: Report to Congressional Committees;* United States Government Accountability Office: Washington, DC, USA, 2018.
2. Wright, T.P. Factors Affecting the Cost of Airplanes. *J. Aeronaut. Sci.* **1936**, *3*, 122–128. [CrossRef]
3. Asher, H. Cost-Quantity Relationships in the Airframe Industry. Ph.D. Thesis, The Ohio State University, Columbus, OH, USA, 1956.
4. Boone, E.R.; Elshaw, J.J.; Koschnick, C.M.; Ritschel, J.D.; Badiru, A.B. A Learning Curve Model Accounting for the Flattening Effect in Production Cycles. *Def. Acquis. Res. J.* **2021**. in-print.
5. Mislick, G.K.; Nussbaum, D.A. *Cost Estimation: Methods and Tools;* John Wiley & Sons: Hoboken, NJ, USA, 2015.
6. Argote, L.; Beckman, S.L.; Epple, D. The Persistence and Transfer of Learning in Industrial Settings. *Manag. Sci.* **1990**, *36*, 140–154. [CrossRef]
7. Argote, L. Group and Organizational Learning Curves: Individual, System and Environmental Components. *Br. J. Soc. Psychol.* **1993**, *32*, 31–51. [CrossRef]
8. Jaber, M.Y. Learning and Forgetting Models and Their Applications. *Handb. Ind. Syst. Eng.* **2006**, *30*, 30–127.
9. Glock, C.H.; Grosse, E.H.; Jaber, M.Y.; Smunt, T.L. Applications of Learning Curves in Production and Operations Management: A Systematic Literature Review. *Comput. Ind. Eng.* **2019**, *131*, 422–441. [CrossRef]
10. Jaber, M.Y.; Bonney, M. Production Breaks and the Learning Curve: The Forgetting Phenomenon. *Appl. Math. Model.* **1996**, *2*, 162–169. [CrossRef]
11. Nembhard, D.A.; Uzumeri, M.V. Experiential Learning and Forgetting for Manual and Cognitive Tasks. *Int. J. Ind. Ergon.* **2000**, *25*, 315–326. [CrossRef]
12. Sikström, S.; Jaber, M.Y. The Depletion–Power–Integration–Latency (DPIL) Model of Spaced and Massed Repetition. *Comput. Ind. Eng.* **2012**, *63*, 323–337. [CrossRef]
13. Anzanello, M.J.; Fogliatto, F.S. Learning Curve Models and Applications: Literature Review and Research Directions. *Int. J. Ind. Ergon.* **2011**, *41*, 573–583. [CrossRef]
14. Crossman, E.R. A Theory of the Acquisition of Speed-Skill*. *Ergonomics* **1959**, *2*, 153–166. [CrossRef]
15. Moore, J.R.; Elshaw, J.J.; Badiru, A.B.; Ritschel, J.D. *Acquisition Challenge: The Importance of Incompressibility in Comparing Learning Curve Models;* US Air Force Cost Analysis Agency: Arlington, TX, USA, 2015.
16. Honious, C.; Johnson, B.; Elshaw, J.; Badiru, A. *The Impact of Learning Curve Model Selection and Criteria for Cost Estimation Accuracy in the DoD;* Air Force Institute of Technology: Wright Patterson AFB, OH, USA, 2016.
17. Baloff, N. Startups in Machine-Intensive Production Systems. *J. Ind. Eng.* **1966**, *17*, 25.
18. Baloff, N. Startup Management. *IEEE Trans. Eng. Manag.* **1970**, *4*, 132–141. [CrossRef]
19. Corlett, E.N.; Morecombe, V.J. Straightening Out Learning Curves. *Pers. Manag.* **1970**, *2*, 14–19.
20. Yelle, L.E. The Learning Curve: Historical Review and Comprehensive Survey. *Decis. Sci.* **1979**, *10*, 302–328. [CrossRef]
21. Hirschmann, W.B. Profit from the Learning-Curve. *Harv. Bus. Rev.* **1964**, *42*, 125–139.
22. Li, G.; Rajagopalan, S. A Learning Curve Model with Knowledge Depreciation. *Eur. J. Oper. Res.* **1998**, *105*, 143–154. [CrossRef]
23. Chalmers, G.; DeCarteret, N. *Relationship for Determining the Optimum Expansibility of the Elements of a Peacetime Aircraft Procurement Program;* Stanford Research Institute: Menlo Park, CA, USA, 1949.
24. De Jong, J.R. The Effects of Increasing Skill on Cycle Time and Its Consequences for Time Standards. *Ergonomics* **1957**, *1*, 51–60. [CrossRef]
25. Knecht, G.R. Costing, Technological Growth and Generalized Learning Curves. *J. Oper. Res. Soc.* **1974**, *25*, 487–491. [CrossRef]
26. Badiru, A.B. Computational Survey of Univariate and Multivariate Learning Curve Models. *IEEE Trans. Eng. Manag.* **1992**, *39*, 176–188. [CrossRef]
27. Office of the Secretary of Defense. 1921-1 Data Item Description. Available online: https://cade.osd.mil/content/cade/files/csdr/dids/archive/1921-1.DI-FNCL-81566B.pdf (accessed on 29 September 2020).
28. Hu, S.-P.; Smith, A. Accuracy Matters: Selecting a Lot-Based Cost Improvement Curve. *J. Cost Anal. Parametr.* **2013**, *6*, 23–42. [CrossRef]
29. Cohen, J. Quantitative Methods in Psychology. *Nature* **1938**, *141*, 613. [CrossRef]

30. Badiru, A.B. Half-Life Learning Curves in the Defense Acquisition Life Cycle. *Def. Acquis. Res. J.* **2012**, *19*, 283–308.

Article

Application of a Semi-Empirical Dynamic Model to Forecast the Propagation of the COVID-19 Epidemics in Spain

Juan Carlos Mora [1,*], Sandra Pérez [2] and Alla Dvorzhak [1]

1 Department of Environment, CIEMAT, Avenida Complutense, 40, 28040 Madrid, Spain; alla.dvorzhak@ciemat.es

2 Sercomex Pharma, C/ Pollensa, 2, 28232 Las Rozas de Madrid, Spain; sandra.perez@sercomex.es

* Correspondence: jc.mora@ciemat.es; Tel.: +34-91-346-6751

Received: 9 September 2020; Accepted: 16 October 2020; Published: 22 October 2020

Abstract: A semiempirical model, based in the logistic map, was developed to forecast the different phases of the COVID-19 epidemic. This paper shows the mathematical model and a proposal for its calibration. Specific results are shown for Spain. Four phases were considered: non-controlled evolution; total lock-down; partial easing of the lock-down; and a phased lock-down easing. For no control the model predicted the infection of a 25% of the Spanish population, 1 million would need intensive care and 700,000 direct deaths. For total lock-down the model predicted 194,000 symptomatic infected, 85,700 hospitalized, 8600 patients needing an Intensive Care Unit (ICU) and 19,500 deaths. The peak was predicted between the 29 March/3 April. For the third phase, with a daily rate $r = 1.03$, the model predicted 400,000 infections and 46,000 $\pm$ 15,000 deaths. The real r was below 1%, and a revision with updated parameters provided a prediction of 250,000 infected and 29,000 $\pm$ 15,000 deaths. The reported values by the end of May were 282,870 infected and 28,552 deaths. After easing of the lock-down the model predicted that the health system would not saturate if r was kept below 1.02. This model provided good accuracy during epidemics development.

Keywords: semi-empirical model; logistic map; COVID-19; SARS-CoV-2

---

1. Introduction

A new respiratory disease, initially dominated by pneumonia, and caused by a coronavirus, was detected at the province of Hubei, in China, at the end of 2019. It was initially named by the World Health Organization (WHO) as 2019-nCoV [1] and renamed in February 2020 by the International Committee on Virus Taxonomy as Severe Acute Respiratory Syndrome coronavirus 2 (SARS-CoV-2), recognizing it as a sister of the SARS-CoV viruses [2,3]. The same day the WHO [4] named the disease as Coronavirus Disease 2019 (COVID-19).

Many efforts have been made since then to mathematically model the spread of the disease in the whole world and in the different countries where the infection arrived. Modelling the epidemics has many practical uses: preparation of national health systems; make provisions of the necessary sanitary material; predict whether and when a saturation of the health system could occur; when and to what extent Non Pharmaceutical Interventions (NPI) [5] should be applied; predict the day when those countermeasures can be relaxed, and so forth. These theoretical approaches to predict the evolution of epidemics often use compartment models as simple as the SIR model (Susceptible, Infectious and Recovered—sometimes called Removed) [6], but this model can be increased in complexity to include different characteristics of an infectious epidemics. For example the model can include individuals who can infect others, without presenting symptoms, what is known as the SEIR model (Susceptible, Exposed, Infectious and Removed); the model can also assume that people who have recovered from the disease lose the immunity after a given time, and therefore they could be infected again, giving rise to the SEIRS models (Susceptible, Exposed, Infectious, Removed and Susceptible); also the deaths and births can be included for long term epidemics, as is the case in the influenza; and many times compartments to distinguish deaths, recovered, hospitalized, and other situations, are included by using empirical ratios (see for instance References [7,8] for further information).

As in any consolidated scientific field, there is abundant literature describing the different mathematical models which can be applied to different diseases, for different population behaviour, or even to define the optimal control strategies [9–13]. In Reference [14] a review of the models used in the India to forecast the behaviour of the COVID-19 was carried out, including among others: compartments (SIR type) models, ARIMA models, machine learning based approaches and logistic curves. This review remarks the very important discrepancies found between those models in the predictions after lockdowns were applied.

Since the SARS-CoV-2 outbreak many efforts have been carried out to adapt these SIR type compartment models to the behaviour of this particular virus. For example, a conceptual representation of a compartment model for COVID-19 disease's spread, developed by the authors of this paper, is shown in Figure 1, where immunization of recovered is assumed to be lost after a given period of time, as happen in other infectious diseases (typically immunity is lost after less than 12 months in the case of the coronavirus causing common cold).

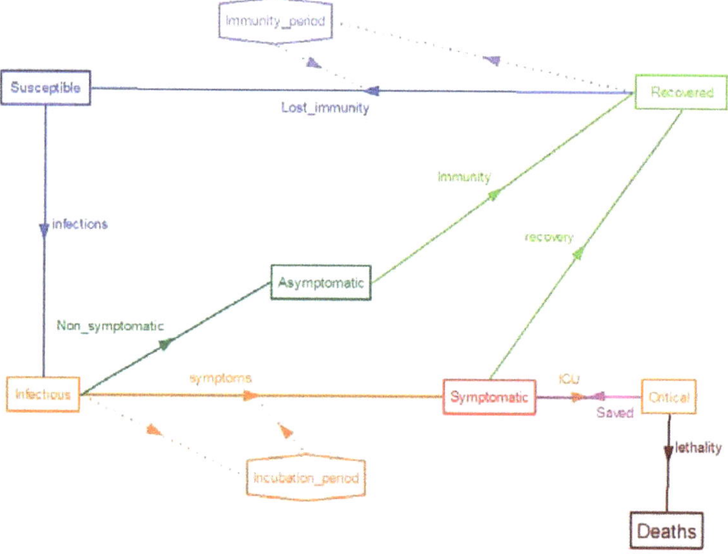

Figure 1. Example of a Susceptible, Infectious and Recovered (SIR) type compartment model adapted to simulate the behaviour of SARS-CoV-2. In this case we assumed that immunity would be lost in a given period of time.

Due to the difficulty of developing and calibrating these compartment models at the early stages of an epidemics, wrong conclusions are often reached, for instance predicting the timing of the epidemic, and many times the uncertainties associated with the results of the models are so wide that are not well accepted by the public. Sometimes the authors of such predictions blamed the quality of the statistical data [15,16]. However, this quality is severely affected by the urgency of the epidemic and could not probably be avoided in this or future outbreaks of epidemics. The continuous publications of medical and epidemiological studies on the COVID-19, and the associated virus, don't make it easy to extract good quality information to adapt the models. But it must be accepted that this situation will be always the case—or even worse—when new diseases appear.

Some other problems associated with future predictions of the COVID-19 behaviour, like the influence of ambient temperature or humidity in the seasonality of the disease [17], are still unsolved. In the early stages many aspects of the behaviour of the SARS-CoV-2 virus has been associated with previous studies on similar viruses as the SARS virus, as the resilience in fomites [18] or the immunization of patients recovered from it [19]. Many other aspects which would also affect the choice of the best model are still under investigation, for example the possible lost of the immunity after some months, as happens with other human coronaviruses which cause other diseases like the 15% of the common cold cases [20].

During the outbreak of the epidemic in Spain several models were tested and a follow-up of the published results were performed to support Spanish national authorities in the decision-making process [21], producing a preliminary work covering all the phases which was published as a preprint [22]. The best results were obtained by using a semi-empirical approach presented herein, which has the advantage of performing accurate predictions with the minimum amount of information available during this epidemic which, very likely, will be the situation in future outbreaks.

This paper presents the mathematical development for a proposed semi-empirical model and the results obtained using it, focusing the results into the Spanish case. Some results obtained for other countries are also presented.

2. Materials and Methods

The semi-empirical model presented in this paper, with a proper calibration, produces accurate results at every stage of the epidemic: during the first spread of the disease, after the application of NPI (specifically total lock-down) which were applied in many countries, and after the ease of the lock-down.

Although the instant reproduction number (Rt) used by the epidemiologists for estimating the severity and evolution of an epidemic [23] is not used in this model, the basic reproduction number (R0) was derived for 10 different countries, ranging from 2.0 to 9.3, which is in a good agreement with previous estimations [24]. The R0 derived from real data are found in Table 1, giving an average of 5.8 ± 2.4, a value almost doubling early estimations [25].

Table 1. Basic Reproductive Number calculated for some studied countries.

Country	R0
France	9.3
USA	8.2
Slovenia	7.5
Norway	6.9
Italy	6.7
UK	6.7
Spain	4.6
Belgium	3.5
South Korea	2.8
Germany	2.0

The model proposed in this paper applies the well-known logistic map, often used for describing the growth of populations and mentioned as an example of chaotic behaviour. This chaotic behaviour depends on a single parameter r (Figure 2 shows a fractal created with the logistic map as a function of r). In this equation values of $r < 1$ would make an epidemic to extinguish. Any r greater than 1 but below 3 will provide an equilibrium in the size of the population for the long term, while values of r higher than 3.56995 would produce a chaotic behaviour on the size of the population.

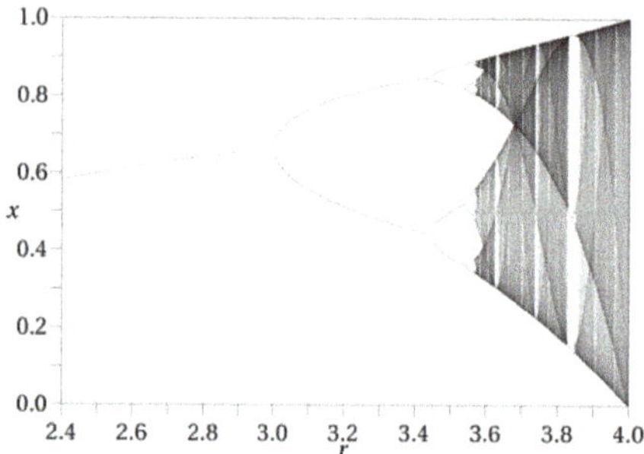

Figure 2. Bifurcation diagram for the logistic map as a function of r.

Therefore to determine the number of infected diagnosed cases Equation (1) is used:

$$I(t) = r \cdot \left(1 - \frac{I(t-1)}{N}\right) \cdot I(t-1), \tag{1}$$

where $I(t)$ is the number of infected diagnosed cases at day t, $I(t-1)$ the infected diagnosed cases of the precedent day $t-1$, r is the growth parameter of the logistic map (named hereafter daily infection rate), and N the number of individuals susceptible to be infected (in Figure 2 a simplified example of this function, with constant parameters, is shown). It should be noted that the number of susceptible individuals used here is not necessarily the same as the number used for modellers applying SIR type models. The sub-index n will be used below to indicate the n-th day after the outbreak.

The behaviour of this function gives rise to the logistic function and the typical sigmoid shape of its cumulative distribution if $r < 2$, while it shows a chaotic behaviour if $r > 3.56995$ (see Figure 3). Other authors have studied the behaviour of this logistic function applied to the COVID-19 epidemics [26,27].

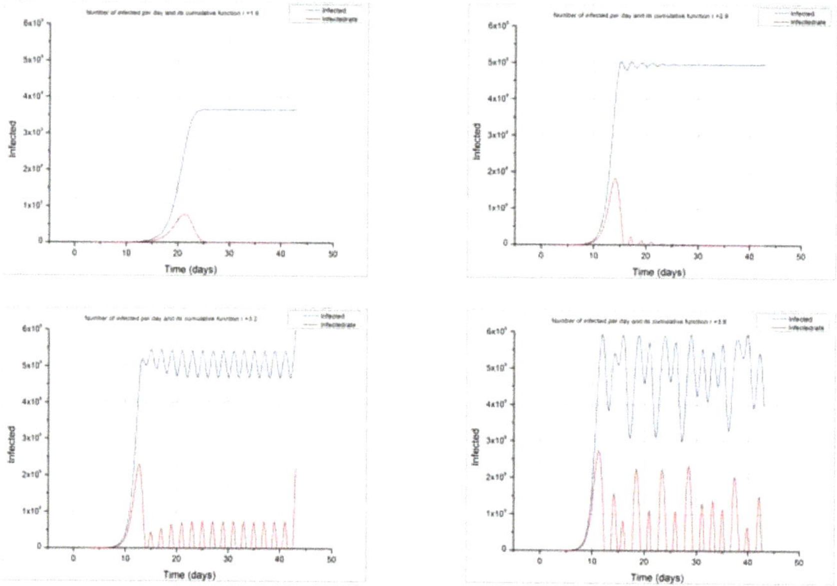

Figure 3. Number of infected obtained for the logistic map as a function of r for the different options, from $r = 1.9$ to $r = 3.8$.

In order to compare with the values of the basic reproduction number in Table 2 the empirically determined values of the growth parameter r are shown for the same countries in Table 1. This r parameter is simply measured by dividing the new infected in a given day by the infected in the previous day. To avoid statistics biases r was taken for each country as an average for the first 7 days after the initial detections of infected at each country. The r in these countries was equal to 1.9 ± 0.5, ranging from 1.3 to 3.0. In this approach a value of $r < 3$ implies that, in absence of countermeasures, and independently of the initial value $I(0)$ there would be reached an unique equilibrium on the number of infected: $I(\infty) = N \cdot \frac{r-1}{r}$. Worldwide, in average, an equilibrium value of 3.65 billions of infected would be reached applying $r = 1.9$ and $N = 7.7 \times 10^9$ to the equation. The equilibrium values which would be reached, if no intervention was applied for each country, are shown at the Table 2.

Table 2. Growth parameter r for the logistic map, empirically calculated for the same countries during the first days of the spread of the COVID-19 epidemics, and equilibrium value for the infected people if no countermeasures were applied.

Country	r	$I(\infty)$ (Millions)
France	1.8	29.7
USA	1.3	75.7
Slovenia	2.2	1.1
Norway	2.0	2.7
Italy	3.0	40.2
UK	1.4	19.0
Spain	1.5	15.5
Belgium	1.7	3.8
South Korea	2.0	51.6
Germany	1.8	36.9

Therefore the basic quantity used to make predictions is the number of infected $I(t)$ reported by each country or region. This model does not need considering asymptomatic infected or questions what is the real number of infected, but makes use of the data reported. However, as demonstrated in the case of the "Diamond Princess" cruise, nearly the 70% of the infected would be asymptomatic and undiagnosed [28].

Other quantities needed to provide advice to the authorities are the number of inpatients who would need medical attention at the hospital (H_n), the number of those who would need intensive care (C_n) and the number of deaths (D_n), all of them at each time t. H_n and D_n are calculated as a fraction of the number of the diagnosed infected cases at time t (I_n), and C_n as fraction of H_n. Obviously the number of recovered patients (R_n) is given by the fraction (1 D_n). The fraction used to calculate D_n in this way is referred to as the case fatality rate (CFR), determined as $CFR = \frac{D_n}{D_n+R_n}$. This is proved to be more practical during the outbreak than other approaches as the mortality rate for the whole population which can be only experimentally known at the end of the epidemic.

A delay must be included to represent the time elapsed between a death and its report to the authorities, including the time needed to perform the diagnoses (usually by using the polymerase chain reaction (PCR) technique).

All of these numbers are crucial to policy makers in order to take well founded decisions. However, to perform reliable predictions an appropriate calibration of the model is needed which will depend on the specific situation of each phase of the epidemics.

2.1. Initial Parametrization

All the parameters of the model are empirically calibrated by averaging the available information in a studied region. This calibration is feasible at the early stages, when the data available cover only few days, but it can also be dynamically adjusted during the whole evolution of the epidemics.

For performing reliable predictions at the very beginning of the outbreak, the information from previously infected countries can be used as initial calibration of the model.

SARS-CoV-2 is assumed to infect with the same probability to every human, disregarding sex or age. Being a new human virus, no immunization was previously acquired, by natural or artificial (vaccine) means. For that reason the number of people which can be infected, N, was initially assumed to be the whole population of the studied region, whatever the size of that region is. For the sake of simplicity the total population of a country (or a region, as are the so-called autonomous communities in Spain) is initially used. In the case of Spain the total population $N = 46.6 \times 10^6$ was initially used.

The daily infection rate, r, can be dynamically determined - using all the data collected to average a given period - dividing the number of infected the day $n + 1$ (I_{n+1}) by the number of infected at day n (I_n). For Spain, averaging the daily infection rate during the first 7 days of the outbreak, from the 26 February/3 March, an $r = 1.5$ was obtained. Following the same method r values were obtained for 10 countries (see Table 2). This parameter however with the actions taken by the governments and the population, as the social distancing, the frequent hands washing or the use of masks.

The fraction of the infected who need to be hospitalized ($H = \frac{H_n}{I_n}$) is dynamically determined using the data acquired at each region (or state, or country), averaged for the whole period since the beginning of the epidemic. The same was done for the fraction of inpatients needing an ICU ($C = \frac{C_n}{Hn}$). An initial factor of patients needing an Intensive Care Unit (ICU) from the number of infected cases $H \cdot C = 0.05$ to 0.15 was computed from the studies in Eastern countries.

As the reported data for the infected patients were given as accumulated since the beginning of the epidemic, all the other quantities were also obtained as cumulative functions. Due to its special configuration of the health system, this was a problem in Spain, as different regions decided to report different quantities and then initially was impossible to obtain accumulated data for hospitalized persons, or patients which needed an ICU for the country. Daily rates were interesting to calculate, for instance, the day where the maximum of infections or deaths (peak of the curve) would occur.

For the CFR, the value measured near the equilibrium in China was used ($CFR = 0.0578$, as measured the 4 March—see Figure 4 to see the evolution of the CFR in China). This parameter presented a similar behaviour in many other countries, reaching a value at equilibrium of nearly a 5%. The high CFR values observed at the beginning of the epidemics in every country are probably due to several joint factors, including the weakness of the more vulnerable population (very old, already sick, people), or the lack of knowledge on which medical treatments were more effective. Those factors improved with time. In the cases of European countries a similar evolution was observed, although the decrease was slower than happened in China (at the beginning of April 2020 the CFR for USA was 0.3998, for Italy was 0.3557, and for Spain was 0.2052). The time delay, from the diagnose of an infected patient to the possible death, was adjusted at each country using real data. In the initial stages the observed delay was of 5 to 10 days for all the countries and reduced after some days.

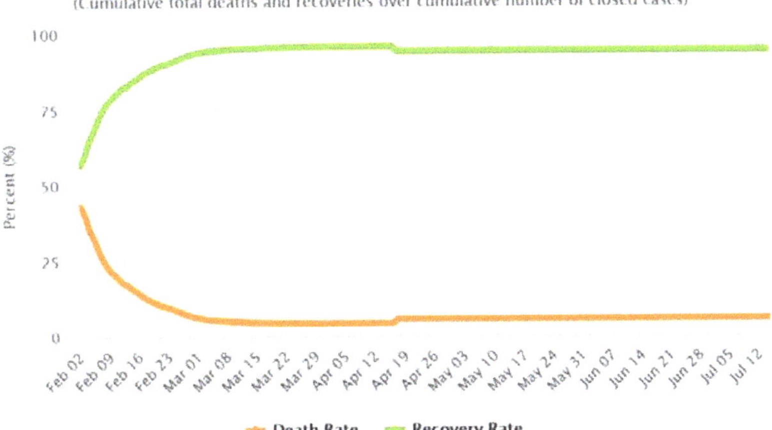

Figure 4. Case Fatality Rate as measured in China since the beginning of February 2020. The value measured the 4th of March of 5.78% was taken for the model (Source of information https://www.worldometers.info/coronavirus/ (consulted on 11 March)).

2.2. *Parametrization during the Lock-Down*

A non pharmaceutical intervention used in China and many other countries, included Spain, was the so called 'lock-down' in which the population is required to stay at home and only leave if essential. This NPI has been partially implemented in some regions and totally in others, including the region of Hubei in China (58.5 million inhabitants), Spain (46.6 million inhabitants) or Italy (60.4 million inhabitants).

In each region or country the initial value used for N was its total population, but after the lock-down, the number of people already infected, or in contact with infectious people, is fixed and therefore N would be smaller. This number cannot be determined before the lock-down but can be calculated the same day that the lock-down is implemented by using the number of infected measured at that exact time. A first estimation was made using the number of infected, estimated with the model, 14 days after beginning the lock-down (14 days was assumed to be the incubation period for the COVID-19) and multiplying that number by a factor of 10, which would provide the total infected. This method provides a rough estimation which needs further refinements when new data are obtained, however it provided valid estimations for forecasting the time when the maximum (peak) for daily new cases of diagnosed cases or deaths would be expected.

As expected, the daily infection rate r was observed to decrease, from the rate the day before the NPI was applied (typically around 1.3) to a number slightly higher than 1.0, as observed at South Korea. The same behaviour was observed in every country and at every scale. After the lock-down is implemented, the r parameter can be fitted by least squares to the curve given in Equation (2), for the given region or country.

$$r = 1 + A \cdot e^{-\alpha \cdot t}, \tag{2}$$

where t is the time in days since the lock-down and A and α are constants empirically determined at the location. Table 3 shows some values for A and α adjusted for different countries after a lock-down was implemented and examples of smaller scale regions within Spain (Andalusia and Catalonia were chosen for this example because they are the two more populated regions in Spain). Those values were obtained by fitting the Equation (2) to the experimental data in different regions or countries. (* Experimental data from worldometer (Source of information https://www.worldometers.info/coro

navirus/ (consulted on 17 April)). ** Experimental data from the Spanish official source of information (Source of information 'Instituto de Salud Carlos III' (ISCIII): https://covid19.isciii.es/)).

Table 3. Values obtained by fitting Equation (2) to the experimental data in different countries and two Spanish regions (Catalonia and Andalusia).

Region	A	α
South Korea *	0.226	0.235
Italy *	0.293	0.070
UK *	0.326	0.051
Spain **	0.295	0.074
Andalusia **	0.366	0.096
Catalonia **	0.491	0.109

* (Source of information https://www.worldometers.info/coronavirus/ (consulted on 17 April)); ** ('Instituto de Salud Carlos III' (ISCIII): https://covid19.isciii.es/(consulted on 17 April)).

The number of individuals infected before the lock-down (N), and the constants A and α cannot be determined prior to the lock-down, as different groups of individuals or societies behave differently under the same exact government instructions, and also different governments provided slightly different instructions. So the only chance to obtain good predictions after the lock-down is to wait for several days to obtain experimental data to be used to fit the curve under the Equation (2). It should be also pointed out that, some sources of information provided data shifted in time or simply just different and consequently the fitting could provide different values for the parameters.

The parameters H and C were determined by averaging the empirical values from the studied region. In Spain the values obtained as an average up to the 6 April were $H = 0.467$ and $C = 0.1497$, which indicates that a high rate of diagnosed infected needed to be hospitalized, or more likely, that only severe cases were diagnosed at the hospitals, needing half of them to be admitted. Also a high percentage of the inpatients (almost a 15%) needed intensive care using ventilators, which was in agreement with the observed pattern in China and other countries. In this case $H \cdot C = 0.07$ (7%) which was in a good agreement with the initial range observed for the inpatients needing an ICU. As all parameters were dynamically calculated every day, the predictions were slightly calibrated daily.

Also for the CFR empirical values were used, as the equilibrium value taken from China was well surpassed in the initial stages in many European countries, although it tended to the same equilibrium value (nearly 5%). Although initially it reached values of even a 50%, the experimental CFR in Spain, as in UK, Belgium, or Italy, was 0.12 (12%) by April. The delay applied from reporting the positive diagnose of a patient to the death (when produced) was reduced to 3 days.

2.3. *Parametrization after Easing the Lock-Down*

When a region or a country decides to relax the confinement, the parameters need a new calibration to take into account the situation.

When the lock-down is completely abandoned N would return to be again the whole population of the region or country. However, this was not the situation in every country.

For example, in Spain the lock-down was established 15 March. Although the ideal situation would be to maintain the total lock-down until r reached a value close to 1, value expected to happen by the end of April according with the model, 13 April some relaxation was adopted, allowing most of the workers to return to their normal activities. A large part of the population remained confined, but a graded approach was established to remove it before the end of May. This being the situation, the parameters can be only inferred after some data are collected, following the same methodology established during the lock-down. Therefore r should be fitted to an exponential decrease, following the same Equation (2) after obtaining enough data. To perform initial conservative estimations a value of $r = 1.03$ can be used.

$$r = 1.01 + B \cdot e^{-\beta \cdot t}. \tag{3}$$

In the final stages of the easing of the lock-down, Equation (3) was used for r, considering impossible to achieve a value $r < 1.01$ (as the experience in other countries showed that reducing the level of daily infections below that value was, at least, very difficult). B and β are again constants empirically determined at the location

The rest of parameters: H, C or the CFR remain being averaged along the whole period with real data.

This parametrization can be used to assess the evolution of the situation after easing the lock-down or to design the strategy to optimize the number of infected, hospitalized or inpatients needing an ICU, to avoid the saturation of the health system of a country.

3. Results

As pointed out, 3 phases were considered:

1. An initial phase of the outbreak where no severe restrictions were applied.
2. A second phase where severe non pharmaceutical interventions (confinement) were applied.
3. A last phase where relaxation of the more severe NPI is assumed, although some keep being used.

As an example the application of the model with the appropriate parametrization in Spain is presented to show the performance of the model. The same methodology was also used for some of the regions in Spain, and can be used to any other region or country of any size.

3.1. Initial Phase

The initial phase is considered before any countermeasure is applied. Figure 5 shows the cumulated number of infected and the total number of deaths reported in Spain (in red) from 29 February/20 March. From 29 February/14 March reported data (in red) are shown against modelled values (in green). The schools closing was established from the 11th of March and the total lock-down the 15th of March.

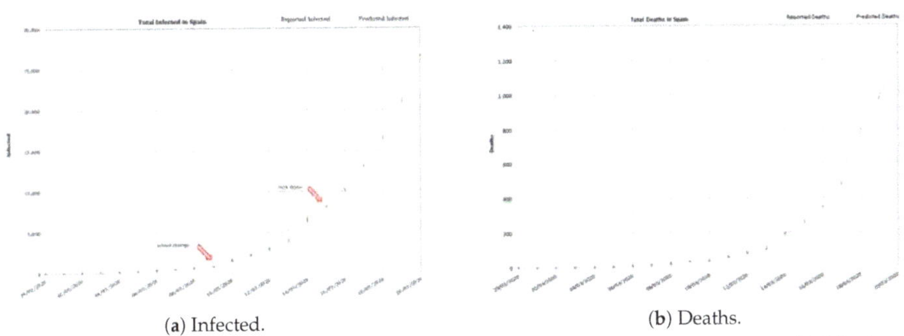

(**a**) Infected. (**b**) Deaths.

Figure 5. Predicted number of infected (**a**) and deaths (**b**) in Spain from 29 February/20 March. The data from the 14 March are based in the model assuming no interventions were implemented. Red dots—reported. Green bars—modelled with uncertainty.

As explained, during this initial stage, $r = 1.05$ was calculated as an average of the values measured during the initial days of spread of the epidemics; N was the total population in Spain (46.6×10^6 inhabitants); and the $CFR = 0.0578$ was taken from the Chinese experience. In this phase the only parameter dynamically calibrated, to adjust the data reported daily, was the delay from the number of infected to the number of deaths, as was explained before, from an initial value of 7 days that was reduced up to a 2 days delay applied the 5th of March. In this specific case, the forecast indicated a number of infected cases of 26,600 $\pm$ 500 and a number of deaths of 1230 $\pm$ 150 to occur 6 days later (11th of March). The real number of reported infected was 21,571 (19% difference), and the number of reported deaths was of 1093 (11% difference). The number of inpatients needing an ICU and a ventilator was calculated as $I(t) \cdot H(t) \cdot C(t)$, providing a range of [1330–3990]. The reported number of inpatients which needed an ICU the 20th of March was of 1630 (within the calculated range). Although in this initial stage many factors altered the real numbers the accuracy was reasonably good.

Predictions of the likely number of infected, hospitalized inpatients and total deaths were carried out using these conditions for the initial phase (uncontrolled spread of the disease). The model predicted that, if a severe NPI (total lock-down) was not adopted, but on the contrary the virus was left to spread without control, at the end of the epidemic in Spain 12 million people would have been infected, of which nearly 1 million people would need intensive care and about 700,000 infected would die directly because of the COVID-19 disease. However the number of deaths would likely be higher due to the saturation of the health system, as these numbers would occur in a very short period.

These results could provide an early idea of the urgent necessity of applying extreme NPIs like the total lock-down, they could be also used to predict the consequences of not applying the severe NPIs, and also to prepare for the capabilities of the ICUs, including the number of ventilators.

3.2. Lock-Down Phase

In Spain closing of schools began on 11 March and lock-down was established the 15 March. All factors were re-calibrated for this second phase as stated, including a fitting of the daily infection rate to the curve given in Equation (2). For the number of susceptible individuals N an initial estimation ($N = 1.1 \times 10^6$) was carried out using the results of the model 14 days after the lock-down. An average value $r = 1.32$ was used. Results of these early predictions are presented in Figure 6. These values were later calibrated dynamically to $r = 1.20$ and $N \simeq 9 \times 10^6$ using data up to the 15th of March. Results of the predictions are presented in Figure 7.

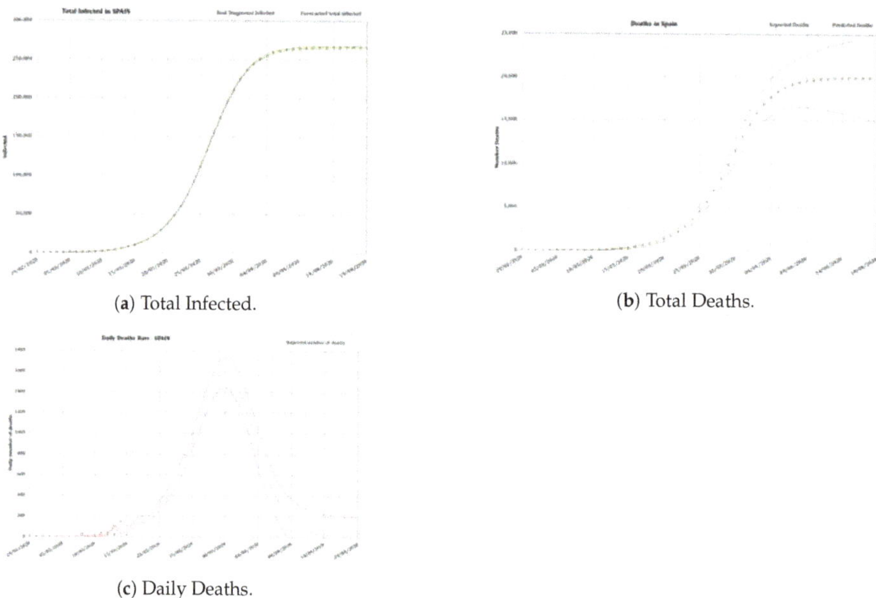

(**a**) Total Infected.

(**b**) Total Deaths.

(**c**) Daily Deaths.

Figure 6. Total number of infected (**a**), total deaths (**b**) and daily deaths (**c**) in Spain predicted from 29 February/19 April. The preliminary results assumed the total lock-down since the 15 March with data up to the 14 March. Red dots—reported data. Green bars—modelled values with uncertainty.

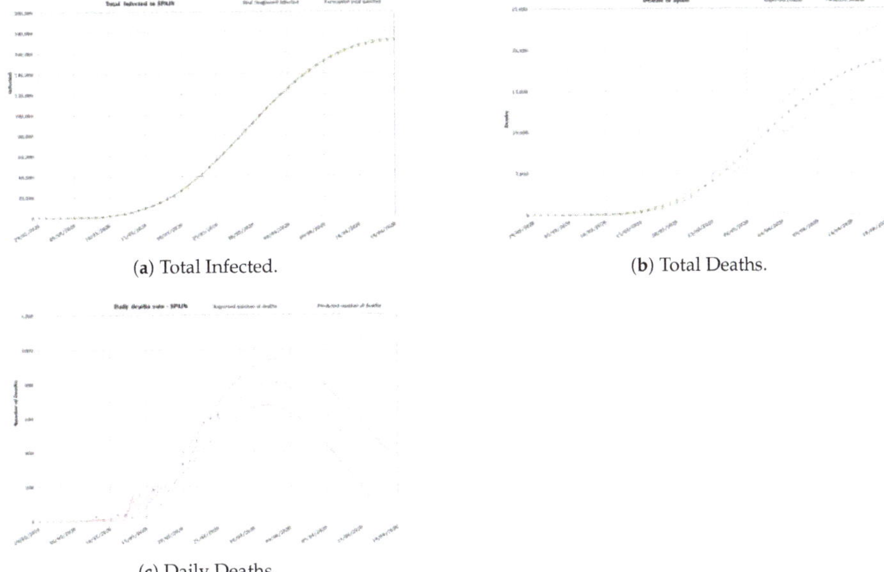

(**a**) Total Infected.

(**b**) Total Deaths.

(**c**) Daily Deaths.

Figure 7. Total number of infected **(a)**, total deaths **(b)** and daily deaths **(c)** in Spain predicted from the 29 February/19 April. These second modelled results assumed the total lock-down since the 15 March and were calibrated with data up to the 25 March. Red dots—reported data. Green bars—modelled values with uncertainty.

The application of the total lock-down to the model reduced the predictions carried out the 26 March to a total of 194,000 infected, 85,700 hospitalized, nearly 8,600 with needs of an ICU and $19{,}500 \pm 1400$ deaths to occur by the 17 April. The real numbers reported at that date were 197,142 infected (1.6% difference), 7548 inpatients needing an ICU (12% difference) and 20,043 deaths (2.7% difference). The model predicted the peak for the rate of daily deaths to occur between the 29th of March and the 3rd of April. In reality the peak, after the administration revised the data (two months later), occurred the 31 March.

3.3. *Unlocking Phase*

The last phase is the ease of the lock-down. In this case, predictions based on some hypothesis, carried out during April, are presented in this paper and compared with the real evolution. Again the case of Spain is presented as example. Using the models presented in this paper recommendations were provided to ease the lock-down around 21 April [21]. In reality a partial unlock was decreed the 13th of April for non-essential workers, and a phased total unlocking from the 30 April, where some activities were allowed gradually each week until the 21 June, where normal activity was restored, although the population should follow NPI health countermeasures, as social distance, use of masks, washing of hands, and so forth.

After the unlocking many uncertainties appear, but the results of the model depend largely on the daily rate of infections r.

3.3.1. Partial Unlock

On the 13 April a partial ease of the total lock-down was applied in Spain.

In order to obtain conservative figures an initial forecast was carried out with the data available the 16 April [22]. At that point only some reasonable hypotheses could be applied to calibrate N and r.

For N it was assumed that about a 20% of the total workforce (in Spain 20 million workers on 2020) went back to work, as only some industries were allowed to begin again their activities, after that day. As those people could also infect their families, 2 further members on average, an initial $N = 1.2 \times 10^7$ was taken. In order to carry out conservative predictions $r = 1.03$ (a daily increase of the infected of a 3%) was taken. Results in Figure 8 were obtained for those conservative assumptions.

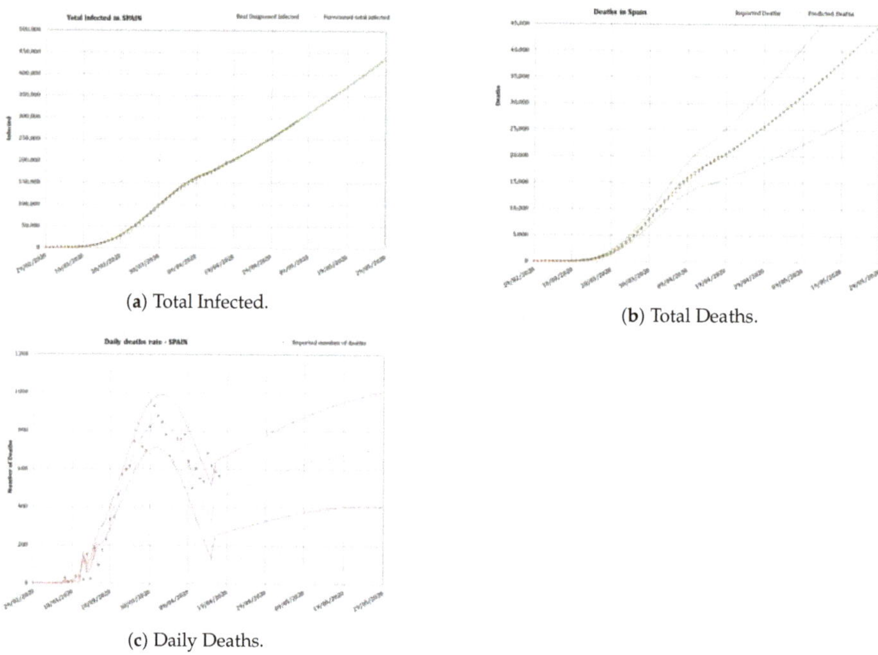

(a) Total Infected.

(b) Total Deaths.

(c) Daily Deaths.

Figure 8. Total number of infected **(a)**, total deaths **(b)** and daily deaths **(c)** in Spain predicted from 29 February/29 May. These modelled results assumed the ease of the lock-down since the 13 April with conservative assumptions for N and r. Red dots—reported data. Green bars—modelled values with uncertainty.

Using these conservative values of N and r, some consequences in the partial ease of the lock-down could be extracted. First of all, the number of diagnosed infected people would increase continuously beyond May. In fact there would been a peak in the daily rate of infected by the 28th of May and a peak in the daily rate of deaths by the 1st of June. The total number of deaths in Spain by the end of April would reach the 46,000 $\pm$ 15,000 under this scenario. That was the conservative value we published in April [22]. If this would have been the case ($r > 1.03$) a saturation of the health system would have occurred again in Spain. So that value of r could be regarded as an upper bound which should be avoided.

In reality, a very good behaviour of the Spanish population made r to continuously reduce even after a total easing of the lock-down.

3.3.2. Total Unlocking

From the 4th of May a total unlock was applied in Spain, with a progressive increase in the mobility of the people since then, and therefore a recalibration was needed. For this phase N was again considered the total population (46.6 millions inhabitants).

Obtaining some more data a least squares fit was performed using an initial daily infection rate $r = 1.02$ (2% daily increase of infected) the day before the unlocking, and the Equation (3) to fit r.

The calibration resulted in $B = 0.03$ and $\beta = 0.153$, what gives the equation $r = 1.01 + 0.03 \cdot e^{(-0.153 \cdot t)}$. The results obtained with this fit is shown in Figure 9.

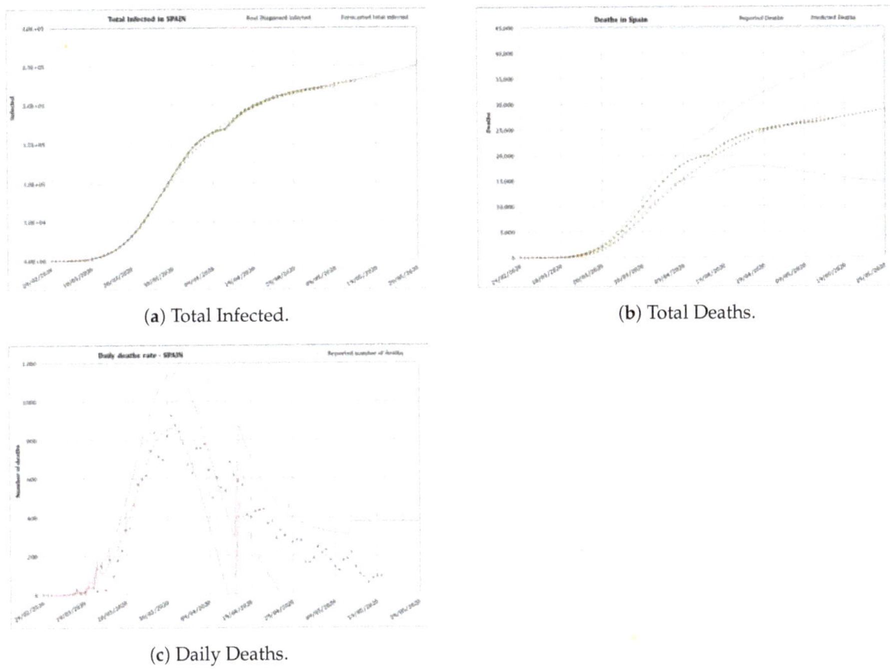

(**a**) Total Infected.

(**b**) Total Deaths.

(**c**) Daily Deaths.

Figure 9. Total number of infected (**a**), total deaths (**b**) and daily deaths (**c**) in Spain predicted from 29 February/28 June. These modelled results assumed the ease of the lock-down since the 13 April. Red dots—reported data. Green bars—modelled values with uncertainties.

The series of data finishes at the end of May, as official aggregated data for Spain were no longer provided. In fact least squares fit was extremely difficult as there was an attempt to homogenize the data between the different regions in the country which made the whole series of data to be revised almost every day. That is one of the reasons why in this occasion the fitting was not as good as in previous phases for the total number of infected and the total number of deaths, as the focus was put in the daily number of deaths to obtain a good fitting. Daily number of deaths was the main endpoint in this phase because this indicator is the best one to perform future surveys of the situation of the epidemic.

The results obtained using this calibration for this final stage was that the number of diagnosed infected people would slowly increase continuously beyond May. This is a logical result, as the infection would be always present in a slow rate until the virus is eradicated or there is a vaccine to control the spread of the disease. The predicted number of total infected is of $317{,}500 \pm 1700$ and the total number of deaths would be 37,100, with a huge uncertainty, by the 1 August. The real numbers that day were 335,602 reported infected (5.7% difference) and 28,445 deaths (23% difference).

During the summer the situation was controlled, however if the good practices in the application of NPIs are abandoned: hands washing, social distance, use of masks, and so forth, r could easily reach values above 1.03, surely another increase in the number of infected will occur. This will be the case also when borders are reopened and new infectious people (even asymptomatic) enter inadvertently from countries in the initial phase of the epidemics.

Of course, these predictions did not consider important changes, as the discovery of a vaccine—which seems extremely difficult in a short period of time—or the increase in the temperatures in the summer which could reduce the infectivity of the SARS-CoV-2, or a higher isolation which could reduce the severity of the COVID-19 due to a higher production of vitamin D [29,30], or any other unforeseen circumstance. During that summer time, of course, the treatment of hospitalized patients has improved and therefore a smaller fraction of inpatients need an ICU or even die.

3.4. Percentage of Infected Population in the Regions

There was an additional result extracted from this model. The need to recalibrate it during the locking phase by fitting the parameter N, the number of people susceptible to be infected in that phase, offers the possibility of using that number to infer the percentage of the population in a country or region which could have been infected, for this particular virus, most of them showing no symptoms.

As an example these numbers were extracted for the autonomous communities (administrative regions) in Spain and transformed to three levels of infection (low—below 5% —, medium—from 5% to 10%— and high—above 10%), giving rise to the result shown in Figure 10a.

This qualitative result has an important use for the authorities, as the population which have a high level of infection by the SARS-CoV-2 did probably developed immunity against the virus (at least temporarily as discussed before), and therefore there is no chance for them to be infected again in the close future. And on the contrary, there is a bigger chance of developing further outbreaks of the disease in those regions where the percentage of the infected population was smaller.

The model provided some counter-intuitive results. For example, the capital of Spain, Madrid, presented "medium level" of infection of the population, whereas Catalonia showed "high level", while both, the number of diagnosed cases and the number of deaths in Madrid was higher, and this would imply a higher level of infection in Madrid. However this could be explained in the different criteria followed by the different regions for reporting the numbers. For example Catalonia decided by mid of May to include in the statistics the deaths of people occurred out of the hospitals, while it was unknown if Madrid was including those deaths already in the statistics. The same occurs with the number of diagnosed infected cases, as there is observed an increase in the number of tests performed on people who finally did not need a hospital in Madrid, the percentage in Catalonia of diagnosed population finally needing a hospital was still around 65% by May.

(**a**) Forecasted levels of infection [22].

(**b**) Measured levels of infection [31].

Figure 10. Levels of infection in the population at each region of Spain: calculated in April 2020 by using the model presented in this paper (**a**) and measured at the seroprevalence study ENE-COVID finished in July 2020 (**b**).

These predictions were compared against the extensive statistical program of immunity prevalence carried out on the Spanish population from April to June 2020, published in early July [31] (see Figure 10b). This study measured the percentage of people infected during the epidemic, taking account of asymptomatic infected persons, while the previous reported numbers included just the hospitalized inpatients showing severe symptoms on which a polymerase chain reaction (PCR) test returned a positive result. The results of both studies can be compared in Figure 10.

As can be seen the results provided by the model in April were accurate in most of the regions were low infection occurred, as in the south (Andalucia, Extremadura, Ceuta, Melilla, Canary Islands, Balears Islands, Valencia and Murcia) or north-west (Galicia, Asturias and Cantabria) of Spain. However the model predicted a medium to high infection level at the whole north and north-east, whereas the measured levels of seroprevalence obtained low levels at some regions (La Rioja, Navarra and Basque Country). In general the forecast provided a good general view of the infection level.

4. Discussion and Conclusions

As it is unlikely that a vaccine to the SARS-CoV-2 or a cure of the associated disease COVID-19 is developed in the next months, the only way of reducing the consequences of the epidemic at this moment is an optimum application of the available NPIs.

Traditional epidemiological models like compartment models need good quality data to offer proper predictions. However, during the outbreak of an epidemic often the quality of the data is not the primary concern of the health systems and therefore many epidemiologists were discouraged and abandoned their efforts of offering predictions. This fact, together with the need of fewer data, made it interesting using a semiempirical model based in the logistic map. The application of such methodology to the different phases of the COVID-19 epidemics in Spain was shown. This methodology provided good results in the forecast of the evolution of the disease in every situation.

The use of extreme non-pharmaceutical interventions, such as the total lock-down, showed their effectiveness during the period they were applied. However, easing the countermeasures allow new outbreaks of the infection to appear. This situation forces the need of applying many simultaneous techniques to reduce the effect of the disease if that is the case. One of those techniques could be the application of the methodology described in this paper to provide early alerts of the outbreaks in countries or smaller units of population, allowing an optimization of sanitary resources and reducing the economic and social impact of future NPIs applied locally.

All the data used in the paper were data officially published in real time by the official sources from the Spanish Government [32,33]. As was shown, reasonably accurate results can be produced by using the model presented in this paper to the different phases of an epidemic. In a previous preprint, assuming an infection daily rate r of 3%, a total number of 400 000 diagnosed infected and a total number of 46,000 $\pm$ 15,000 deaths were forecasted in Spain by the end of May [22]. Those predictions overestimated the real values due to a more strict reduction of the infection daily rate in the country, reaching values below 1%.

The forecasts covered from the number of infected, hospitalized, inpatients needing an ICU or deaths, to the time where the peak of daily deaths would be produced or the level of infection in a given region. In the last prediction, carried out for the beginning of August, 317,500 $\pm$ 1700 infected and a total number of deaths of 37,100 were predicted, with a huge uncertainty, to be compared with the real numbers of 335,602 reported infected (5.7% difference) and 28,445 deaths (23% difference).

The aim of any policy dealing with the application and withdrawal of NPI should carefully consider daily infection rates. In the case of the COVID-19 a daily infection rate r lying within the range of 1.01 to 1.02 (1% to 2% daily increase), as was shown in countries like South Korea, would produce a manageable level of people needing an ICU in hospitals, avoiding the saturation of national healthcare systems and therefore unnecessary deaths.

Also a qualitative prediction of the percentage of the population infected in the different regions of Spain was performed by using the suggested semi-empirical model. These predictions were compared against the extensive statistical program of immunity prevalence carried out on the Spanish population from April to June 2020, published at early July, showing that the model provided in April reasonable results in most of the regions, although the model predicted a medium to high infection level at the whole north and north-east, while the measured levels of seroprevalence obtained low levels.

Some results obtained with this methodology were not intuitive according to the official information. The more counter-intuitive result probably being the higher level of infection of Catalonia compared with Madrid region. As said, in general the forecast provided a good forecast of the infection level.

The COVID-19 epidemic is still ongoing and the knowledge will increase with time. In the next future new outbreaks are foreseen in the countries where the first one was controlled, unless a vaccine or a cure are developed in the next future. Therefore models will be needed to forecast again the evolution and to advice the authorities in the needs of the country's health system. Some characteristics of the virus, needed to perform better predictions, are still unknown, as the lost of immunity of cured individuals or the influence of vitamin D in the severity of the disease.

A continuous watch of the disease is still needed to provide proper advice which can be used by policy makers.

Author Contributions: Conceptualization, Data Curation, Methodology, Software and Writing by J.C.M.; Validation, Writing—review and editing and Formal analysis by S.P. and A.D. All authors have read and agreed to the published version of the manuscript.

Funding: This research received no external funding.

Acknowledgments: The authors would like to acknowledge to all the colleagues who have constructively read and discussed the paper to improve its content. Specially to Ignacio Rodríguez and Asunción Núñez from Dragonfly Innovation Technologies, UK; Justin Smith, from the Public Health England, UK; and Alejandro Ubeda from the Ramon y Cajal Hospital, Spain.

Conflicts of Interest: The authors declare no conflict of interest.

Abbreviations

The following abbreviations are used in this manuscript:

CoV	Coronavirus
COVID19	Coronavirus Disease 2019
CFR	Case Fatality Rate
DOAJ	Directory of Open Access Journals
ICU	Intensive Care Units
MDPI	Multidisciplinary Digital Publishing Institute
NPI	Non Pharmaceutical Interventions
SARS	Severe Acute Respiratory Syndrome
SEIR	Susceptible Exposed Infectious Removed
SEIRS	Susceptible, Exposed, Infectious, Removed and Susceptible
SIR	Susceptible Infectious Removed
WHO	World Health Organization

References

1. Zhu, N.; Zhang, D.; Wang, W.; Li, X.; Yang, B.; Song, J.; Zhao, X.; Huang, B.; Shi, W.; Lu, R.; et al. A Novel Coronavirus from Patients with Pneumonia in China, 2019. *N. Engl. J. Med.* **2020**, *382*, 727–733. [CrossRef] [PubMed]
2. Gorbalenya, A.E.; Baker, S.C.; Baric, R.S.; de Groot, R.J.; Drosten, C.; Gulyaeva, A.A.; Haagmans, B.L.; Lauber, C.; Leontovich, A.M.; Neuman, B.W.; et al. Severe acute respiratory syndrome-related coronavirus: The species and its viruses—A statement of the Coronavirus Study Group. *bioRxiv* **2020**. Available online: https://www.biorxiv.org/content/early/2020/02/11/2020.02.07.937862.full.pdf (accessed on 2 April 2020). [CrossRef]
3. Gorbalenya, A.E.; Baker, S.C.; Baric, R.S.; de Groot, R.J.; Drosten, C.; Gulyaeva, A.A.; Haagmans, B.L.; Lauber, C.; Leontovich, A.M.; Neuman, B.W.; et al. The species Severe acute respiratory syndrome-related coronavirus: Classifying 2019-nCoV and naming it SARS-CoV-2. *Nat. Microbiol.* **2020**, *5*, 536–544. [CrossRef]
4. World Health Organization. Coronavirus Disease. Available online: https://www.who.int/emergencies/diseases/novel-coronavirus-2019 (accessed on 2 April 2020).

5. Feng, L.; Kumar, M.; Mark, L. An optimal control theory approach to non-pharmaceutical interventions. *BMC Infect. Diseases* **2010**, *10*. [CrossRef]
6. Kermack, W.O.; McKendrick, A.G.; Walker, G.T. A contribution to the mathematical theory of epidemics. *Proc. R. Soc. Lond.* **1927**, *115*, 700–721. [CrossRef]
7. Brauer, F. *Compartmental Models in Epidemiology.*; Springer: Berlin/Heidelberg, Germany, 2008; pp. 19–79. [CrossRef]
8. Munz, P.; Hudea, I.; Imad, J.; Smith, R.J. *When Zombies Attack!: Mathematical Modelling of an Outbreak of Zombie Infection*; Tchuenche, J.M., Chiyaka, C., Eds.; Infectious Disease Modelling Research Progress; Nova Science Publishers, Inc.: Hauppauge, NY, USA, 2009; pp. 133–150.
9. Daley, D.J.; Gani, J. *Epidemic Modelling: An Introduction*; Cambridge University Press: Cambridge, UK, 1999; Chapter 15, pp. 379–404. [CrossRef]
10. Choisy, M.; Guégan, J.F.; Rohani, P. *Mathematical Modeling of Infectious Diseases Dynamics*; John Wiley and Sons, Ltd.: Hoboken, NJ, USA, 2006; Chapter 22, pp. 379–404. Available online: https://onlinelibrary.wiley.com/doi/pdf/10.1002/9780470114209.ch22 (accessed on 2 October 2020). [CrossRef]
11. Martcheva, M. *An Introduction to Mathematical Epidemiology*; Springer: Berlin/Heidelberg, Germany, 2015. [CrossRef]
12. Li, M.Y. *An Introduction to Mathematical Modeling of Infectious Diseases*; Springer: Berlin/Heidelberg, Germany, 2018. [CrossRef]
13. Brauer, F.; Castillo-Chavez, C.; Feng, Z. *Mathematical Models in Epidemiology*; Springer: Berlin/Heidelberg, Germany, 2019. [CrossRef]
14. Arora, S.; Jain, R.; Singh, H. Epidemiological Models of SARS-CoV-2 (COVID-19) to Control the Transmission Based on Current Evidence: A Systematic Review. *Preprints.org* **2020**. [CrossRef]
15. Caudill, L. Lack of Data Makes Predicting COVID-19's Spread Difficult but Models are Still Vital. Available online: https://theconversation.com/lack-of-data-makes-predicting-covid-19s-spread-difficult-but-models-are-still-vital-135797. (accessed on 15 April 2020).
16. Roda, W.C.; Varughese, M.B.; Han, D.; Li, M.Y. Why is it difficult to accurately predict the COVID-19 epidemic? *Infect. Disease Model.* **2020**, *5*, 271–281. [CrossRef] [PubMed]
17. Wang, J.; Tang, K.; Feng, K.; Lv, W. High Temperature and High Humidity Reduce the Transmission of COVID-19. *SSRN* **2020**. [CrossRef]
18. Kampf, G.; Todt, D.; Pfaender, S.; Steinmann, E. Persistence of coronaviruses on inanimate surfaces and their inactivation with biocidal agents. *J. Hosp. Infect.* **2020**, *104*, 246–251. [CrossRef] [PubMed]
19. Prompetchara, E.; Ketloy, C.; Palaga, T. Immune responses in COVID-19 and potential vaccines: Lessons learned from SARS and MERS epidemic. *Asian Pac. J. Allergy Immunol.* **2020**, *38*, 1–9. [CrossRef] [PubMed]
20. Pelczar, J.R.; Chan, E.; Kieg, N.R. *Microorganism and Disease: Microbial Diseases*; Mc Graw Hill, Tata: Berlin/Heidelberg, Germany, 2010; p. 656.
21. Mora, J.C. *Prediction of the Advance of the SARS-CoV-2 Virus (COVID-19)*; Three reports issued on 15 and 26 March and 7 April 2020; Department of Environment, CIEMAT: Madrid, Spain, 2020.
22. Mora, J.C.; Perez, S.; Rodriguez, I.; Nunez, A.; Dvorzhak, A. A Semiempirical Dynamical Model to Forecast the Propagation of Epidemics: The Case of the SARS-CoV-2 in Spain. *medRxiv* **2020**. Available online: https://www.medrxiv.org/content/early/2020/04/23/2020.04.19.20071860.full.pdf (accessed on 20 April 2020). [CrossRef]
23. Chintalapudi, N.; Battineni, G.; Sagaro, G.G.; Amenta, F. COVID-19 outbreak reproduction number estimations and forecasting in Marche, Italy. *Int. J. Infect. Diseases* **2020**, *96*, 327–333. [CrossRef] [PubMed]
24. Liu, Y.; Gayle, A.A.; Wilder-Smith, A.; Rocklöv, J. The reproductive number of COVID-19 is higher compared to SARS coronavirus. *J. Travel Med.* **2020**, *27*. Available online: https://academic.oup.com/jtm/article-pdf/27/2/taaa021/32902430/taaa021.pdf (accessed on 20 April 2020). [CrossRef] [PubMed]
25. Velavan, T.P.; Meyer, C.G. The COVID-19 epidemic. *Trop. Med. Int. Health* **2020**, *25*, 278–280. [CrossRef] [PubMed]
26. Fokas, A.S.; Dikaios, N.; Kastis, G.A. Predictive mathematical models for the number of individuals infected with COVID-19. *medRxiv* **2020**. Available online: https://www.medrxiv.org/content/early/2020/05/06/2020.05.02.20088591.full.pdf (accessed on 2 October 2020). [CrossRef]

27. Wu, K.; Darcet, D.; Wang, Q.; Sornette, D. Generalized logistic growth modeling of the COVID-19 outbreak in 29 provinces in China and in the rest of the world. *medRxiv* **2020**. Available online: https://www.medrxiv.org/content/early/2020/03/16/2020.03.11.20034363.full.pdf (accessed on 2 October 2020). [CrossRef] [PubMed]
28. Emery, J.C.; Russel, T.W.; Liu, Y.; Hellewell, J.; Pearson, C.A.; Knight, G.M.; Eggo, R.M.; Kucharski, A.J.; Funk, S.; Flasche, S.; et al. The contribution of asymptomatic SARS-CoV-2 infections to transmission—A model-based analysis of the Diamond Princess outbreak. *medRxiv* **2020**. Available online: https://www.medrxiv.org/content/early/2020/05/11/2020.05.07.20093849.full.pdf (accessed on 2 October 2020). [CrossRef]
29. Ilie, P.C.; Stefanescu, S.; Smith, L. The role of vitamin D in the prevention of coronavirus disease 2019 infection and mortality. *Aging Clin. Exp. Res.* **2020**, *32*, 1195–1198. [CrossRef] [PubMed]
30. Panagiotou, G.; Tee, S.A.; Ihsan, Y.; Athar, W.; Marchitelli, G.; Kelly, D.; Boot, C.S.; Stock, N.; Macfarlane, J.; Martineau, A.R.; et al. Low serum 25-hydroxyvitamin D (25[OH]D) levels in patients hospitalised with COVID-19 are associated with greater disease severity. *Clin. Endocrinol.* **2020**. Available online: https://onlinelibrary.wiley.com/doi/pdf/10.1111/cen.14276 (accessed on 2 October 2020). [CrossRef] [PubMed]
31. ISCIII. Estudio ENE-COVID: Informe final estudio nacional de sero-epidemiología de la infección por SARS-COV-2 en España. Available online: https://www.mscbs.gob.es/ciudadanos/ene-covid/docs/ESTUDIO_ENE-COVID19_INFORME_FINAL.pdf (accessed on 2 October 2020).
32. ISCIII. Datos Notificados por las CCAA a la RENAVE. Available online: https://cnecovid.isciii.es/covid19/ (accessed on 2 October 2020).
33. Ministerio de Sanidad de España. Actualizaciones de la enfermedad por SARS-CoV-2 (COVID-19). Available online: https://www.mscbs.gob.es/profesionales/saludPublica/ccayes/alertasActual/nCov/situacionActual.htm (accessed on 2 October 2020).

 *forecasting* 

Article

Photovoltaic Output Power Estimation and Baseline Prediction Approach for a Residential Distribution Network with Behind-the-Meter Systems

Keda Pan [1], Changhong Xie [2], Chun Sing Lai [2,3,*], Dongxiao Wang [2,4] and Loi Lei Lai [1,2,*]

1 Department of Control Engineering, School of Automation, Guangdong University of Technology, Guangzhou 510006, China; 1111904017@mail2.gdut.edu.cn
2 Department of Electrical Engineering, School of Automation, Guangdong University of Technology, Guangzhou 510006, China; 2111904168@mail2.gdut.edu.cn (C.X.); dongxiao.wang@aemo.com.au (D.W.)
3 Brunel Institute of Power Systems, Department of Electronic & Electrical Engineering, Brunel University London, London UB8 3PH, UK
4 System Design and Engineering Department, Australia Energy Market Operator, Melbourne 3000, Australia
* Correspondence: chunsing.lai@brunel.ac.uk (C.S.L.); l.l.lai@ieee.org (L.L.L.)

Received: 6 October 2020; Accepted: 29 October 2020; Published: 2 November 2020

Abstract: Considering that most of the photovoltaic (PV) data are behind-the-meter (BTM), there is a great challenge to implement effective demand response projects and make a precise customer baseline (CBL) prediction. To solve the problem, this paper proposes a data-driven PV output power estimation approach using only net load data, temperature data, and solar irradiation data. We first obtain the relationship between delta actual load and delta temperature by calculating the delta net load from matching the net load of irradiation for an approximate day with the least squares method. Then we match and make a difference of the net load with similar electricity consumption behavior to establish the relationship between delta PV output power and delta irradiation. Finally, we get the PV output power and implement PV-load decoupling by modifying the relationship between delta PV and delta irradiation. The case studies verify the effectiveness of the approach and it provides an important reference to perform PV-load decoupling and CBL prediction in a residential distribution network with BTM PV systems.

Keywords: PV output power estimation; PV-load decoupling; behind-the-meter PV; baseline prediction

1. Introduction

1.1. Background and Motivation

As the disadvantages of fossil fuel power generation become increasingly prominent, renewable energy generation is developing rapidly, especially photovoltaic (PV) generation [1–3]. The installed power capacity of renewable energy generation grew more than 200 GW, which is mostly PV generation in 2019 [4,5]. However, because of the intermittency and uncertainty of PV, the high penetration of PV could bring great challenges to the power grid, such as power distribution system planning and operation [6–9], load demand forecasting [10–12], hybrid energy system configuration [13,14], and PV power forecasting [15,16].

Demand response (DR) [17–19] is an effective method to improve the reliability and flexibility of the power distribution system [20–22]. In order to incent customers to participate in the DR program, the DR aggregators need to make decisions based on the customer baseline (CBL). Therefore, it is vital to predict the CBL accurately [23]. However, in a residential distribution network, the distributed PV owned by customers generally is located behind-the-meter (BTM) [24], which measures the net load and denotes actual electricity load minus PV power generation. Hence, the distributed PV generation, such as rooftop PV, makes it difficult to predict CBL with net load [25] due to the volatility of PV and load.

To solve the above issues, it is necessary to decouple the PV generation and electricity consumption load from net load data. The well-known method is to install additional meters to specifically monitor each PV for operators; however, it is infeasible as a result of high extra cost and privacy issues. Consequently, for a residential distribution network with BTM PV systems, the innovative approach to decouple PV-load and predict CBL from net load data is developed in this paper.

1.2. Literature Review

In recent years, there has been a large amount of literature on BTM PV, which involves BTM PV detection and capacity estimation [24,26], and BTM PV output power prediction [27,28], etc. In the meantime, it has been a heated research topic for PV-load decoupling and CBL forecasting. One of the common approaches is the PV physical model [29,30], established to simulate and disaggregate PV generation power; however, the detailed PV panel parameters (such as the size, material, azimuth, and tilt) and meteorological information including temperature and solar irradiation need to be known. The authors of [31] realized PV-load decoupling and PV parameters estimating through combining an iterative method with models including the PV physical model with the load estimation model. However, its performance particularly depends on the accuracy of the physical model. Fortunately, a variety of advanced measurement facilities have appeared in the power grid with the development of information technology, e.g., supervisory control and data acquisition (SCADA) and smart meters, etc. It has spawned the other approach (a data-driven approach), which has been investigated by some researchers to decouple PV output power from net load data.

The net load measured by smart meters was used to estimate the individual distributed PV's capacity and generation power in [15] through a support vector regression model. Based on the hybrid data dimension reduction method and mapping function, the authors of [32] provided a data-driven method to predict BTM PV generation power. However, they need to meter the output power of a small amount of BTM PV output power, which is assumed to have the similar characteristics as the estimated BTM PV. In [33], the unsupervised method disaggregating PV was proposed, considering customers' similar electricity consumption behavior with and without PV. Then, the authors in [34] further developed a disaggregation model to decouple PV-load considering a battery energy storage system (BESS). Nevertheless, the authors of [35] showed that it is different for the user with PV and without PV to consume electricity. To avoid these issues, considering both the PV output power and the harmonic current injected by the inverter of PV, the method was carried out in [36] to calculate the PV generation power. However, this method may not be appropriate when some PV installed without permission are included in the power distribution system.

In addition, the baseline (BL) prediction has been studied in some works. In [37], an improved support vector regression model with the ambient temperature of two hours before a DR event as input variables was implemented to predict the office building BL. The authors of [38] studied the CBL impacts on the profit of the company and customers through obtaining the CBL prediction of residential customers, and building the cost and profit function. To address the non-synchronous matching issues in the previous study, the authors in [39] proposed a clustering-based method to build a prediction model considered only on past DR daily load data instead of non-DR daily load data. The authors of [40] proposed a two-stage adaptive BL prediction method that combined the self-organizing map and k-means clustering methods to identify days similar to the tested day under DR events. Considering that determined BL prediction cannot reflect the users' complex electricity consumption behavior, in [41] the authors proposed a probabilistic BL prediction model based on Gaussian process regression. In [42], customers are selected randomly to form a control group to predict the CBL.

Reviewing the above articles, most CBL predictions were for users who had not installed a distributed PV system. However, considering the impact of BTM PV, all separated PV output power information in the residential distribution network is unknown except for the net load data and meteorology data, which will greatly affect the final prediction accuracy.

To address the above issues in this paper, based on the nearest neighbor algorithm and artificial neural network method, a PV-load decoupling and CBL prediction approach is proposed for a residential power distribution system with BTM PV. The main contributions of this paper are as below:

1. The net load is decoupled from PV output power and actual load precisely;
2. To correct the deviation of the matched net load data, the relationship between PV output power and the solar irradiance, and the relationship between actual load and the temperature, are discovered and further formulated;
3. The CBL is predicted based on the PV-load decoupling.

This paper is organized as follows. Section 2 introduces the problem formulation. The methods are provided in Section 3, including the data set division, the electricity consumption sensitivity analysis, PV output power sensitivity analysis, and baseline prediction. In Section 4, the case studies are shown. Finally, the conclusions are given in Section 5.

2. Problem Formulation

In the residential distribution network with BTM PV, both PV output power and actual load are concealed in the net load. It brings a huge challenge for operators to make some decisions including optimal power dispatching and BL prediction.

2.1. PV Output Decoupling

Same as the customer net load, the aggregated net load is composed of the aggregated actual load and aggregated PV output power. Let $D = \{d|1, 2, \ldots, D_D\}$ be a set of day record and $T = \{t|1, 2, \ldots, T_T\}$ be a set of timestamp, then the aggregated net load on time t day d can be formulated as follows:

$$P_{nl}(d,t) = P_{ac}(d,t) - P_{PV}(d,t), \forall d \in D, \forall t \in T \tag{1}$$

where $P_{nl}(d,t)$, $P_{ac}(d,t)$, and $P_{PV}(d,t)$ are the aggregated net load, aggregated actual load, and aggregated PV output power, respectively. Figure 1 gives an example for the composition of net load.

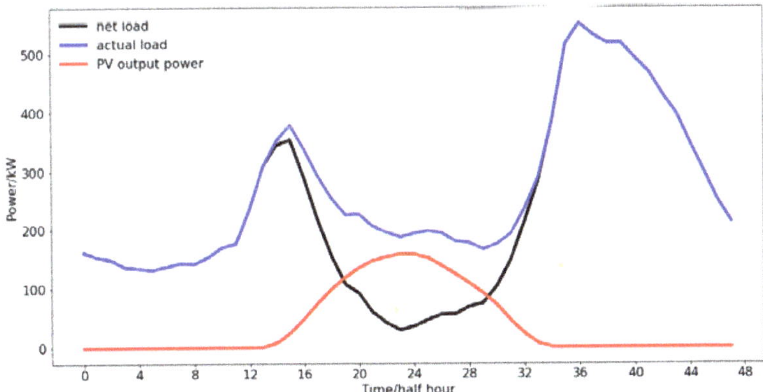

Figure 1. Example of the composition of net load.

Considering that the roof solar PV systems are usually invisible for the aggregator, and in most cases, the small scale distributed PV systems (less than 10 kW) are not set up with a separated PV output power meter [24], it is a big challenge to obtain the pure PV output power without the relevant monitoring information. Therefore, the purpose of the study is to acquire the decomposed PV output power by historical load data, solar irradiation data, and temperature data.

Unlike industrial load, temperature is the main impact factor for residential load consumption behavior [43]. As for the magnitude of PV output, the solar irradiation received by the PV panel plays a decisive role [44]. Based on the above-mentioned characteristics of user electricity consumption and PV output power, this paper considers the sensitivity of actual load to temperature to correct the difference in actual load due to temperature in net load on different days, and then the solar irradiation and PV power relationship is obtained.

2.2. CBL Prediction

For users who have not installed distributed PV equipment, relatively few factors can affect the CBL prediction. The error of the CBL prediction mainly comes from the diversity of the electricity consumption behavior of aggregated users under different environmental conditions [38]. However, for the users with PV systems, the uncertainty of PV output power should also be taken into account. To reduce the error of forecasting the CBL when only net load is available, in this paper, the CBL of actual load is predicted initially based on the PV decomposition technology, and then combined with the PV output power to improve the CBL prediction results of aggregated users equipped with distributed PV equipment.

3. Methods

3.1. Data Set Division

Due to the large differences in solar irradiation and temperature amplitude in different seasons, it leads to different electricity consumption patterns and changes in PV output power. The data set is divided into four categories according to local seasonal conditions. The detailed division time of four categories are shown in Table 1.

Table 1. Division time of four categories.

Season	Label	Duration
Spring	1	Early September to end of October
Summer	2	Early November to end of March
Autumn	3	Early April to end of May
Winter	4	Early June to end of August

Since the PV equipment does not generate electricity before sunrise and after sunset, a period $\tau = \{t_{rise}, t_{rise}+1, \ldots, t_{set}\}$, $\tau \in T$ is set to indicate the time containing solar irradiation, where t_{rise} and t_{set} represent the time of sunrise and time of sunset, respectively. The net load data of each season is further divided into the part that contains PV power, named $P_{nl1}(s,d,t)$, and the part that does not contain PV power, named $P_{nl2}(s,d,t)$, where $s \in 1,2,3,4$, represents the label of seasons.

3.2. Load Consumption Sensitivity Analysis

3.2.1. Correlation between Electricity Consumption and Temperature

Aggregated consumers have different electricity consumption habits in different seasons, and the difference in electricity consumption behavior is largely driven by temperature. In order to reflect the degree of correlation between temperature and consumer electricity consumption, a statistical method called Pearson's correlation coefficient [41] is used to measure the strength of the relationship between the two variables, as shown in Equation (2).

$$r(T_{m,av}(d), P_{ac,av}(d)) = \frac{\sum_{d=1}^{H}\left(T_{m,av}(d) - \overline{T}_{m,av}\right)\left(P_{ac,av}(d) - \overline{P}_{ac,av}\right)}{\sqrt{\sum_{d=1}^{H}\left(T_{m,av}(d) - \overline{T}_{m,av}\right)^2}\sqrt{\sum_{d=1}^{H}\left(P_{ac,av}(d) - \overline{P}_{ac,av}\right)^2}} \quad (2)$$

where $T_{m,av}$ and $P_{ac,av}$ represent the average temperature and average actual load, and $\overline{T}_{m,av}$ and $\overline{P}_{ac,av}$ represent the average value of T days average daily temperature and average actual load. The absolute values of Pearson's correlation coefficient are less than or equal to 1. The closer the r value is to 1 or −1, the stronger the positive or negative correlation between the two calculated variables. Conversely, the closer the r value is to 0, the weaker the correlation between the calculated two variables.

3.2.2. Electricity Consumption Sensitivity Model

Although temperature is the main factor that affects actual load, it is almost impossible to get the numerical relationship between temperature and actual load, because actual load consumption is invisible in the considered scenario. What is certain is that the PV output power is similar when the solar irradiation is the same or very similar [26,27] at the same time in the same season. Thus, if we can find a suitable match of pairs of similar or consistent solar irradiation dates and record them, and further make a difference to the corresponding net load data from the recorded date, we can eliminate the impact of PV output data as much as possible and get the actual load difference as formulated by the following equation:

$$\begin{aligned} &P_{nl1}(s, d^i, t) - P_{nl1}(s, d^j, t) \approx P_{ac1}(s, d^i, t) - P_{ac1}(s, d^j, t) \\ &s.t.\ R(s, d^i, t) \approx R(s, d^j, t) \end{aligned} \quad (3)$$

where R denotes the solar irradiation, and P_{ac1} denotes the actual load during the time period with solar irradiation. Since it is almost impossible to find a completely consistent daily PV curve pairing, we choose to search for a match at each time $t \in \tau$. Considering that for a single point in time, the approximate date of illumination is generally more than one, and to ensure the number of matching samples, for each time t, we use the absolute difference between two points as the evaluation criterion of similarity and select h dates with the most similar solar irradiation data to match. With $\Delta R^h_{rad}\left(d^i, t\right)$ denoting the set of solar irradiance difference between day d^i and another day with similar solar irradiance, $\left|\Delta R\left(d^{i,h}_{rad}, t\right)\right|$ is the absolute difference values between the solar irradiation of day i and the solar irradiation of its hth closest day, and $I^h_{rad}\left(d^i, t\right)$ is the corresponding recording set of the date of solar irradiation similarity match. Their relationship could be given as follows:

$$\begin{cases} \Delta R^h_{rad}\left(d^i, t\right) = \left\{\left|\Delta R\left(d^{i,1}_{rad}, t\right)\right|, \left|\Delta R\left(d^{i,2}_{rad}, t\right)\right|, \ldots, \left|\Delta R\left(d^{i,h}_{rad}, t\right)\right|\right\} \\ \Delta R\left(d^{i,j}_{rad}, t\right) = R\left(s, d^i, t\right) - R\left(s, d^j, t\right) \end{cases} \quad (4)$$

$$I^h_{rad}\left(d^i, t\right) = \left\{d^{i,1}_{rad}, d^{i,2}_{rad}, \ldots d^{i,h}_{rad}\right\} \quad (5)$$

The delta temperature set $\Delta T^h_{m,rad}\left(d^i, t\right)$ and approximate delta actual load set $\Delta P^{h*}_{ac,rad}\left(d^i, t\right)$ according to the date match $I^h_{rad}\left(d^i, t\right)$ can be calculated by Equations (6) and (7), respectively, according to the data match in Equation (5).

$$\begin{cases} \Delta T^h_{m,rad}\left(d^i, t\right) = \left\{\Delta T_m\left(d^{i,1}_{rad}, t\right), \Delta T_m\left(d^{i,2}_{rad}, t\right), \ldots, \Delta T_m\left(d^{i,h}_{rad}, t\right)\right\} \\ \Delta T_m\left(d^{i,h}_{rad}, t\right) = T_m\left(d^i_{rad}, t\right) - T_m\left(d^h_{rad}, t\right) \end{cases}, t \in \tau, i \neq h \quad (6)$$

$$\begin{cases} \Delta P^{h*}_{ac,rad}\left(d^i,t\right) = \left\{\Delta P^*_{ac}\left(d^{i,1}_{rad},t\right), \Delta P^*_{ac}\left(d^{i,2}_{rad},t\right), \ldots, \Delta P^*_{ac}\left(d^{i,h}_{rad},t\right)\right\} \\ \Delta P^*_{ac}\left(d^{i,h}_{rad},t\right) = P_{nl}\left(d^i_{rad},t\right) - P_{nl}\left(d^h_{rad},t\right) \end{cases}, t \in \tau, i \neq h \tag{7}$$

where $\Delta T_m\left(d^{i,h}_{rad},t\right)$ represents the temperature difference between $T_m\left(d^i_{rad},t\right)$ and $T_m\left(d^h_{rad},t\right)$. $\Delta P^*_{ac}\left(d^{i,h}_{rad},t\right)$ is the approximate actual load difference between $P_{nl}\left(d^i_{rad},t\right)$ and $P_{nl}\left(d^h_{rad},t\right)$.

Given the delta temperature set and the approximate delta actual load set, the aim of this part is to perform a fit between these two sets of variables. From Section 3.2.1 we know that the temperature and actual load have a strong linear correlation. Especially in summer and winter, showing strong positive and negative correlations, respectively. Thus, we consider using a linear regression for curve fitting [45]. While taking into account that in the hypothesis when in the case of the same temperature situation, the actual load of the users should also be the same, that is to say when $\Delta T_m\left(d^{i,h}_{rad},t\right)$ equals to 0, $\Delta P^*_{ac}\left(d^{i,h}_{rad},t\right)$ also equals to 0. Therefore, the proportional function is selected as the fitting function and the least squares method is used to solve slope $k_{temp,ac}(t)$, which can be expressed in Equation (8).

$$\begin{cases} \Delta P^{h*}_{ac,rad}\left(d^i,t\right) = k_{temp,ac}(t) \times \Delta T^h_{m,rad}\left(d^i,t\right) \\ k_{temp,ac}(t) = \dfrac{\sum\limits_{i=1}^{n_s}\sum\limits_{h} \Delta T_m\left(d^{i,h}_{rad},t\right) \times \Delta P^*_{ac}\left(d^{i,h}_{rad},t\right)}{\sum\limits_{i=1}^{n_s}\sum\limits_{k} \Delta T_m\left(d^{i,h}_{rad},t\right)^2} \end{cases} \tag{8}$$

where n_s represents the total number of days in season s. Noting that each time $t \in \tau$ has a slope $k_{temp,ac}(t)$, hence there are $(t_s - t_r + 1)$ slopes in total. Figure 2 shows the example of scatter plots of delta temperature and delta actual load at 17:00 in winter and summer and its corresponding linear fitting line.

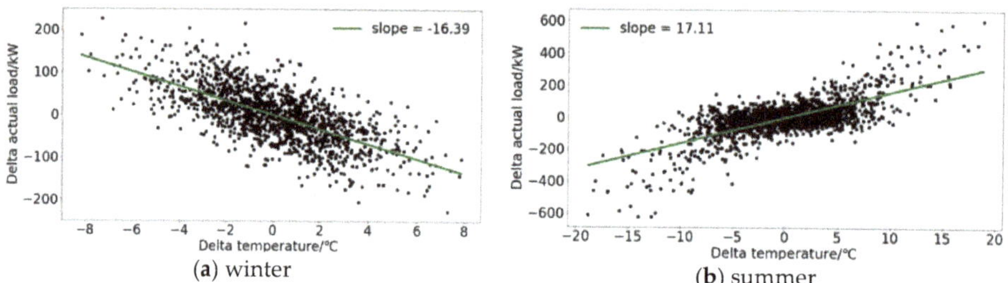

Figure 2. An example of the relationship between delta temperature and delta actual load at 17:00 in winter and summer.

3.3. PV Output Power Sensitivity Analysis

3.3.1. PV Output Power Sensitivity Model

Similar to the method mentioned in Section 3.2.2, if a similar actual load consumption behavior in two days is found, then we make a difference of net load of the corresponding days, and the interrelated approximated delta PV output power can be obtained. Even though the total actual load cannot be found, partial actual load can still be determined. Because of the absence of sunlight where the period in which PV outputs equals to zero, the net load is given as:

$$P_{nl2}(d^i,t) = P_{ac2}(d^i,t),\ t \in \overline{\tau}, \tau \cup \overline{\tau} = T \tag{9}$$

where $\overline{\tau}$ is the complement of τ.

Generally speaking, users' electricity consumption behavior is continuous. Therefore, if the daily electricity consumption behavior in $t \in \overline{\tau}$ period is as similar as possible, the electricity consumption behavior in $t \in \overline{\tau}$ period is as similar as possible to a large extent. The k nearest neighbor (KNN) algorithm [46] is used to find the approximate P_{nl2}.

K nearest neighbor is one of the most simple and effective data mining algorithms. It is proposed based on similar sample points close to each other in space. The principle of the KNN classification algorithm is that if a sample to be classified has k most similar (i.e., the nearest neighbors in the feature space) samples in the feature space, most of which belong to a certain category, the sample also belongs to this category. Therefore, when applying KNN to search for similar electricity consumption behavior curves, the similarity measure is performed on the feature space of each day d^i, and the first k approximate samples are selected as the k neighbors of d^i. The feature space of d^i is $P_{nl2}\left(d^i,t\right)$, $t \in \overline{\tau}$ and the similarity of feature space of d^i and d^j is assessed via Euclidean distance $\rho\left(d^{i,j}\right)$ using Equation (10).

$$\rho\left(d^{i,j}\right) = \sqrt{\sum_{t\in\overline{\tau}}\left(P_{nl2}(d^i,t) - P_{nl2}\left(d^j,t\right)\right)^2}, i \neq j \tag{10}$$

The smaller the value of $\rho\left(d^{i,j}\right)$, the more similar the electricity consumption behavior in the two days (i.e., day i and j). Hence, we can also select k dates with the most similar electricity consumption behaviors data to match. The dates recording set is as follows:

$$I^k_{beh}\left(d^i\right) = \left\{d^{i,1}_{beh}, d^{i,2}_{beh}, \ldots d^{i,k}_{beh}\right\}, i \neq k \tag{11}$$

where $d^{i,k}_{beh}$ is the recording date of electricity consumption behavior of kth nearest neighbor. Then we can make the following assumptions:

$$\begin{array}{l} P_{nl1}(d^i,t) - P_{nl1}(d^j,t) \approx P_{PV1}(d^j,t) - P_{PV1}(d^i,t) \\ s.t.\ \rho\left(d^{i,j}\right) \approx 0 \end{array} \tag{12}$$

The delta solar irradiation set $\Delta R^k_{beh}\left(d^i,t\right)$ and approximate delta PV output power set $\Delta P^{k*}_{PV,beh}\left(d^i,t\right)$ can be calculated by Equations (13) and (14), according to the dates recording in Equations (11) and (12).

$$\left\{\begin{array}{l} \Delta R^k_{beh}\left(d^i,t\right) = \left\{\Delta R\left(d^{i,1}_{beh},t\right), \Delta R\left(d^{i,2}_{beh},t\right), \cdots, \Delta R\left(d^{i,k}_{beh},t\right)\right\} \\ \Delta R\left(d^{i,k}_{beh},t\right) = R\left(d^i_{beh},t\right) - R\left(d^k_{beh},t\right) \end{array}\right., t \in \tau, i \neq k \tag{13}$$

$$\left\{\begin{array}{l} \Delta P^{k*}_{PV,beh}\left(d^i,t\right) = \left\{\Delta P^*_{PV}\left(d^{i,1}_{beh},t\right), \Delta P^*_{PV}\left(d^{i,2}_{beh},t\right), \ldots, \Delta P^*_{PV}\left(d^{i,k}_{beh},t\right)\right\} \\ \Delta P^*_{PV}\left(d^{i,k}_{beh},t\right) = P_{nl}\left(d^i_{beh},t\right) - P_{nl}\left(d^k_{beh},t\right) \end{array}\right., t \in \tau, i \neq k \tag{14}$$

where $\Delta R\left(d^{i,k}_{beh},t\right)$ represents the solar irradiation difference between $R\left(d^i_{beh},t\right)$ and $R\left(d^k_{beh},t\right)$. $\Delta P^*_{PV}\left(d^{i,k}_{beh},t\right)$ is the approximate PV output power difference between $P_{nl}\left(d^i_{beh},t\right)$ and $P_{nl}\left(d^k_{beh},t\right)$.

Now we have the two sequences $\Delta R^k_{beh}\left(d^i,t\right)$ and $\Delta P^{k*}_{PV,beh}\left(d^i,t\right)$, the next step is to get the relationship between these two sequences. It is worth noting that, unlike the electricity consumption sensitivity model mentioned in Section 3.3.1, when the temperature is equal to 0, the electricity consumption is not necessarily 0. For this section, when the solar irradiation is equal to 0, the PV output is also 0. That is to say, through this characteristic, when we find the relationship between delta solar irradiation and approximate delta PV output power, we also find the relationship between solar irradiation and PV output power. The derivation process is formulated as:

$$\left\{\begin{array}{l} \Delta R\left(d^{i,k}_{beh},t\right) = R\left(d^i_{beh},t\right) - R\left(d^k_{beh},t\right) = Rad_{amp}(t) - 0 \\ \Delta P^*_{PV}\left(d^{i,k}_{beh},t\right) = P_{nl}\left(d^i_{beh},t\right) - P_{nl}\left(d^k_{beh},t\right) \approx -P_{PV}\left(d^i_{beh},t\right) + P_{PV}\left(d^k_{beh},t\right) = PV_{amp}(t) - 0 \\ \Delta R\left(d^{i,k}_{beh},t\right) \sim \Delta P^*_{PV}\left(d^{i,k}_{beh},t\right) \Rightarrow Rad_{amp}(t) \sim PV_{amp}(t) \end{array}\right. \tag{15}$$

where $Rad_{amp}(t)$ represents the equivalent solar irradiation amplitude of $\Delta R^{k}_{beh}\left(d^{i},t\right)$, $PV_{amp}(t)$ represents the equivalent PV output power amplitude of $\Delta P^{*}_{PV}\left(d^{i,k}_{beh},t\right)$, and ~ represents the symbol of derivation.

3.3.2. PV Output Power Sensitivity Model Based on Electricity Consumption Sensitivity Correction

The accuracy of decoupling the PV output power curve using the method in Section 3.3.1 depends largely on whether we can find the power consumption curves as similar as possible to match. However, due to actual conditions (insufficient historical data or similar electricity consumption behaviors that have not appeared in historical data), it is nearly impossible to find two completely consistent electrical behavior curves. Therefore, in most cases, there are large errors in the estimation of PV output curves according to the method in Section 3.3.1.

Our mentality is to choose two suitable dates to offset the actual consumption load P_{ac} part of net load P_{nl} as much as possible. However, the similarity of the matching curve in the τ period can only guarantee the similarity of the part of actual load of the matching curve pair in the $\overline{\tau}$ period to a certain extent. Therefore, we hope to make some corrections to modify the two curves to have more similar electricity consumption behavior during the $\overline{\tau}$ period. Although we cannot directly tell what the delta load ΔP_{ac} of the matched two-day curve is in the $\overline{\tau}$ period, we can gain insights about the temperature difference ΔT between the corresponding two days in the $\overline{\tau}$ period. In other words, through the relationship of Equation (8) obtained in Section 3.2.2, we can approximate the correction of actual load difference between the two matching curves P_{cor} in Equation (16).

$$P_{cor}\left(d^{i,k}_{beh},t\right) = k_{temp,ac}(t) \times \Delta T\left(d^{i,k}_{beh},t\right) \tag{16}$$

Then, Equation (15) can be rewritten as:

$$\begin{cases} \Delta R\left(d^{i,k}_{beh},t\right) = R\left(d^{i}_{beh},t\right) - R\left(d^{k}_{beh},t\right) = Rad_{amp}(t) - 0 \\ \Delta P^{*}_{PV}{}'\left(d^{i,k}_{beh},t\right) \ = P_{nl}\left(d^{i}_{beh},t\right) - P_{nl}\left(d^{k}_{beh},t\right) - P_{cor}\left(d^{i,k}_{beh},t\right) \\ \qquad \approx -P_{PV}\left(d^{i}_{beh},t\right) + P_{PV}\left(d^{k}_{beh},t\right) - P_{cor}\left(d^{i,k}_{beh},t\right) = PV'_{amp}(t) - 0 \\ \Delta R\left(d^{i,k}_{beh},t\right) \sim \Delta P^{*}_{PV}{}'\left(d^{i,k}_{beh},t\right) \Rightarrow Rad_{amp}(t) \sim PV'_{amp}(t) \end{cases} \tag{17}$$

where $\Delta P^{*}_{PV}{}'\left(d^{i,k}_{beh},t\right)$ represents the correction of $\Delta P^{*}_{PV}\left(d^{i,k}_{beh},t\right)$, and $PV'_{amp}(t)$ represents the equivalent PV output amplitude of $\Delta P^{*}_{PV}{}'\left(d^{i,k}_{beh},t\right)$.

3.3.3. Evaluation Index of the PV Output Power Estimation

Mean absolute error (MAE) is used as the first evaluation indicator to calculate the accuracy of the decoupling results; the definition formula is as follows:

$$MAE_{PV} = \frac{1}{N}\sum_{t\in\tau}\left|P_{est}(t) - P_{real}(t)\right| \tag{18}$$

where $P_{est}(t)$ and $P_{real}(t)$ represent the estimated value and real value of the PV output power at time t, and N represents the total time stamps during period τ.

Considering the PV output power value is extremely small or zero during the periods of sunrise and sunset or under certain extreme weather conditions, it is therefore not suitable to use mean absolute percentage error (MAPE) as a judgment indicator. However, we still want to show the decomposition effect more intuitively, so the following formula is defined to reflect the relative error of the estimated daily PV output power:

$$RAE_{PV,daily} = \frac{\sum\limits_{t\in\tau}\left|P_{est}(t) - P_{real}(t)\right|}{\sum\limits_{t\in\tau} P_{real}(t)} \tag{19}$$

where $RAE_{PV,daily}$ represents the daily relative absolute error (RAE).

The decomposition process of PV output power from the net load proposed is shown in the following Figure 3.

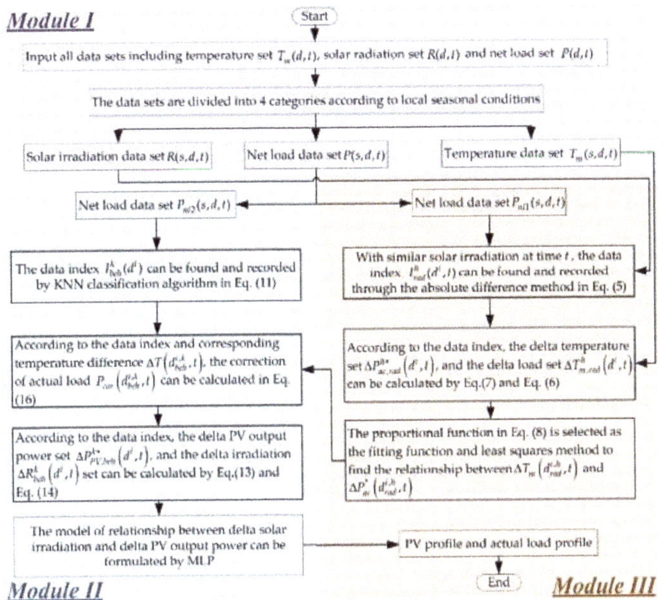

Figure 3. Flowchart of photovoltaic (PV) output power decomposition.

3.4. CBL Prediction Model Construction

In this section, the decomposition technology of PV output power is applied to forecast the CBL of user groups equipped with distributed PV systems to improve the accuracy of the prediction results. Since the CBL prediction makes no difference from the situation without a distributed PV system during the period $\overline{\tau}$, the CBL prediction in this paper mainly focuses on the scenario of the period τ. There are two methods for CBL prediction.

3.4.1. Direct Prediction Method

For the situation of PV output behind-the-meter, the CBL can only be predicted by the history net load of the non-DR event days. For CBL forecasting techniques, the most commonly used and effective method is the averaging algorithms and regression algorithms. The averaging algorithms mainly have the following three types, as given in Table 2.

Table 2. Summary of the averaging methods.

Baseline Estimation Model	Definition
High *X* of *Y*	The average load of the *X* highest consumption days within those *Y* non-DR days preceding the DR event days
Low *X* of *Y*	The average load of the *X* lowest consumption days within those *Y* non-DR days preceding the DR event days
Mid *X* of *Y*	The average load of the *X* middle consumption days within those *Y* non-DR days preceding the DR event days

In this paper, we also use multilayer perceptron (MLP) [47] and recurrent neural network (RNN) [48] as representatives of regression algorithms, and use historical net load data to forecast CBL.

From the definition of each average model and the characteristics of the regression algorithm, CBL of DR days can only be obtained through historical net load in the direct estimation method. This method may increase the forecast error due to the inability to consider the uncertainty of PV.

3.4.2. Prediction Method Based on PV Output Separation

Since we can peel off the PV output power through temperature and solar irradiation data, the actual load and PV output power on DR days can be predicted separately. The CBL prediction of the actual load part $P_{ac,BL}$ can be considered as the CBL prediction from the users that have not installed a distributed PV system. Then, we can get the final CBL prediction $P_{nl,BL}(t)$ by subtracting the part of separated PV output power $P_{PV,BL}(t)$; the equation is given as follows:

$$P_{nl,BL}(t) = P_{ac,BL} - P_{PV,BL}(t). \tag{20}$$

3.4.3. Evaluation Index of CBL Prediction

We evaluate the effect of CBL prediction methods from three perspectives (accuracy, bias, and variability). MAE is used for the evaluation of accuracy, which represents the absolute value of the difference between the true CBL and the predicted CBL. Bias is measured using the mean of the average error between the predicted CBL and the true CBL. Relative error ratio (RER) is used for the evaluation of variability, which represents a fraction of average load during the period τ [39].

4. Case Study

This section shows the effectiveness of the proposed PV output power curve decomposition and its application to CBL prediction.

4.1. Experimental Data Set and Platform Description

We used the load data set sourced from 300 randomly selected solar customers in Ausgrid's electricity network area as the experimental data. The customers chosen had a full set of actual load data and PV output power data for the period from 1 July 2010 to 30 June 2011. In other words, we had intuitive and separate statistics of PV output power data to measure the accuracy of the proposed decomposition method. The temperature and solar irradiation data of the relevant area were from the website [49]. The resolution of the data sets are 48 points per day.

The CPU used in the experiment was an Intel(R) Core(TM) i5-6500 @ 3.20 GHz and the RAM was 8 GB. Python 3.7.7 was used for experimental simulation. The aggregated data was summed by 300 customers' data. Timewise, 7:00–17:30 was set as the τ period. For the PV decomposition part of the experiment, the data of 1096 days in 3 years were used to derive the relationship between the solar irradiation and the PV output power. The total processing times of the method was 43.8 s, which meets the requirements of real-time PV output power estimation.

According to the season division rules in this paper, there are 98 days in spring, 460 days in summer, 169 days in autumn, and 369 days in winter. The clusters of actual load curves of four seasons are shown in Figure 4.

It can be seen from the figure that the actual load is relatively large due to the influence of air conditioning in summer and heating equipment in winter. For spring and autumn, the load has a similar shape. This is because the temperature in Australia in spring and autumn is similar, and the electricity consumption behavior of users is relatively consistent. Although the clusters of various seasons in the figure show that individual curves do not obey the group trends, one of the reasons could be due to extreme weather (cold wave and heat wave) in Australia. The division based on seasons can still effectively aggregate the actual load curves with the same electricity consumption pattern on the whole, which shows that clustering the net load based on different seasons is a good foundation for solving the temperature sensitivity of load consumption in the next step.

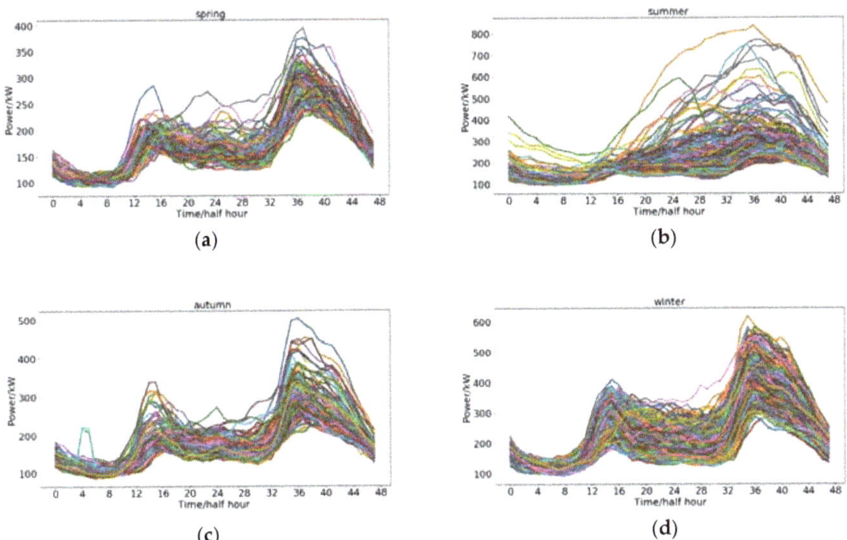

Figure 4. Clusters of actual load curves of (**a**) spring, (**b**) summer, (**c**) autumn and (**d**) winter.

4.2. PV Output Power Curve Separation

The Pearson correlation coefficient between average daily temperature and average daily load for each season are −0.48 (spring), 0.67 (summer), −0.36 (autumn), and −0.74 (winter). In other words, there is a strong correlation between temperature and actual consumption load. Therefore, we used absolute difference to find similar solar irradiation points in each season to match and offset the impact of PV output by making a difference to the net load of the corresponding day, and further used the least squares method to find the relationship between delta temperature and approximate actual load difference of each time stamp. Each solar irradiation point found the five most similar solar irradiation points to match. The slope of each time point of the linear fitting of temperature difference and load difference using the least squares method is shown in Figure 5. It can be observed that the slopes at different times in different seasons are inconsistent, and the positive and negative values of the slopes can also reflect the positive correlation between summer temperature and actual load, and the negative correlation among spring, autumn, and winter.

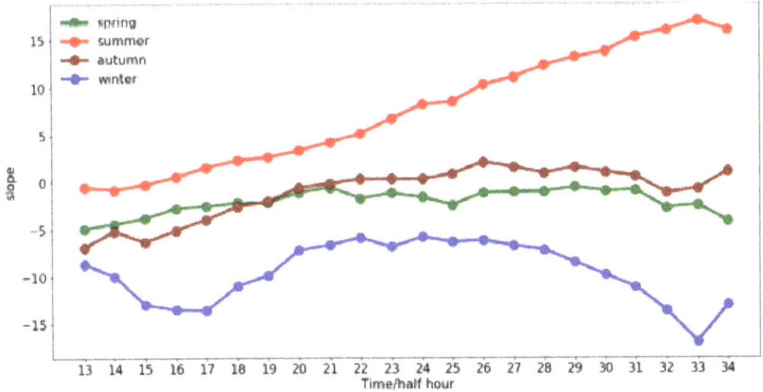

Figure 5. Slope of the linear fitting of delta temperature and approximate delta actual load.

After obtaining the temperature-based sensitivity results of the actual load, the two-day load based on similar electricity consumption behavior can be corrected to the same temperature level according to the corresponding temperature difference (i.e., simulated as a situation where the electricity consumption behavior is as consistent as possible) during the analytic of the PV output power sensitivity.

When analyzing the sensitivity of PV output power, the k value of k nearest neighbor was set to 15 to match load days with similar electricity consumption behavior. We used MLP to build a model of the relationship between delta solar irradiation and approximate delta PV output power. The MLP has a hidden layer with 20 hidden nodes, and the input-hidden layer and hidden-output layer are connected by the rectified linear unit (ReLU) and sigmoid function, respectively. The Adam optimizer was implemented in the MLP. In order to increase the training features of the sample, when building the MLP, in addition to delta solar irradiation, the solar irradiation of the two days before the difference was added as the input of the network.

Table 3 shows the comparison of the separation accuracy of the PV output curve before and after the correction. It can be observed from the table that after the temperature-based correction, the separation accuracy of the PV output power curve has been greatly improved. The value of RAE in spring, summer, autumn, and winter is decreased by 31.60%, 44.08%, 42.40%, and 60.51%, respectively.

Table 3. Accuracy comparison of PV output separation.

Season	Without Temperature Correction		With Temperature Correction	
	MAE (kW)	RAE (%)	MAE (kW)	RAE (%)
Spring	56.47	62.21	27.79	30.61
Summer	70.25	78.42	30.76	34.34
Autumn	51.83	70.66	20.72	28.26
Winter	61.68	98.96	23.97	38.45

In order to more intuitively reflect the effectiveness of the proposed PV output decomposition method, Figure 6 shows the decomposition results of PV output power for a week in each season. It can be seen from the figure that without temperature-based correction, the estimated PV output power will appear negative, which is obviously not logical. The main reason for this problem is that, although we have used KNN to find dates with similar temperatures as much as possible to ensure that the electricity consumption behavior of users on the two days is consistent, however, in reality, it is almost impossible to find a pair of dates with exactly the same temperature change. Moreover, for periods of sunrise and sunset, the solar zenith angle is too small, which leads to very high sensitivity of PV output. When using neural networks to build PV output power estimation models, they are extremely susceptible to differences in electrical behavior caused by temperature differences. Therefore, the PV output power and solar irradiation will show a negative correlation. However, after the temperature-based correction, the negative value is eliminated, and there is also a certain improvement in the curve waveform.

4.3. Results and Analysis of CBL Prediction

Since none of the customers in the data set participated in the DR project, we chose the top 50 electricity consumption days in the data set as the DR days, because the DR event generally occurs when the electricity consumption is large, and the true value of CBL can be regarded as the historical net load value of the corresponding date. At the same time, we could not determine the specific time when the DR event occurred. We assumed the scenario of the DR event occurring at the moment when there was PV output power to reveal the effectiveness of the CBL prediction algorithm based on PV output separation.

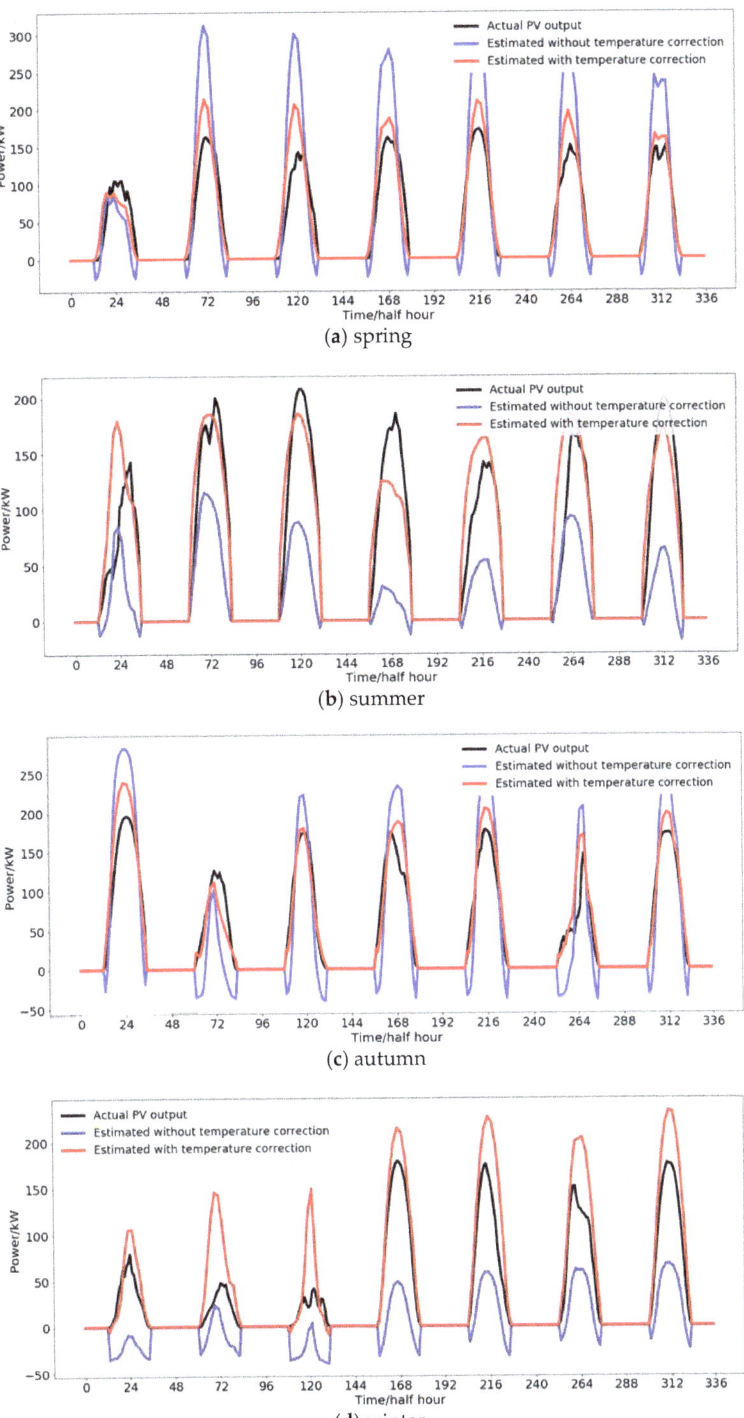

(**a**) spring

(**b**) summer

(**c**) autumn

(**d**) winter

Figure 6. Estimated PV output power curves for a week.

The number of hidden layer units of MLP and RNN were set to 20, and the number of layers was set to 2. We performed 500 iterations for each network training. Historical load and temperature data of 30 days before DR days were used to train these two networks.

The comparison forecasting results of the five methods (High 5 of 10, Low 5 of 10, Mid 5 of 10, MLP, and RNN) of the CBL based on direct prediction and PV curve decomposition are shown in Figure 6, in which a certain DR event day is randomly selected as an example. In Figure 7, we can see that after using the proposed PV output power curve decomposition algorithm, the accuracy of all five CBL prediction methods has been improved. Table 4 shows the average results of the evaluation indexes of CBL prediction for overall DR days under different methods.

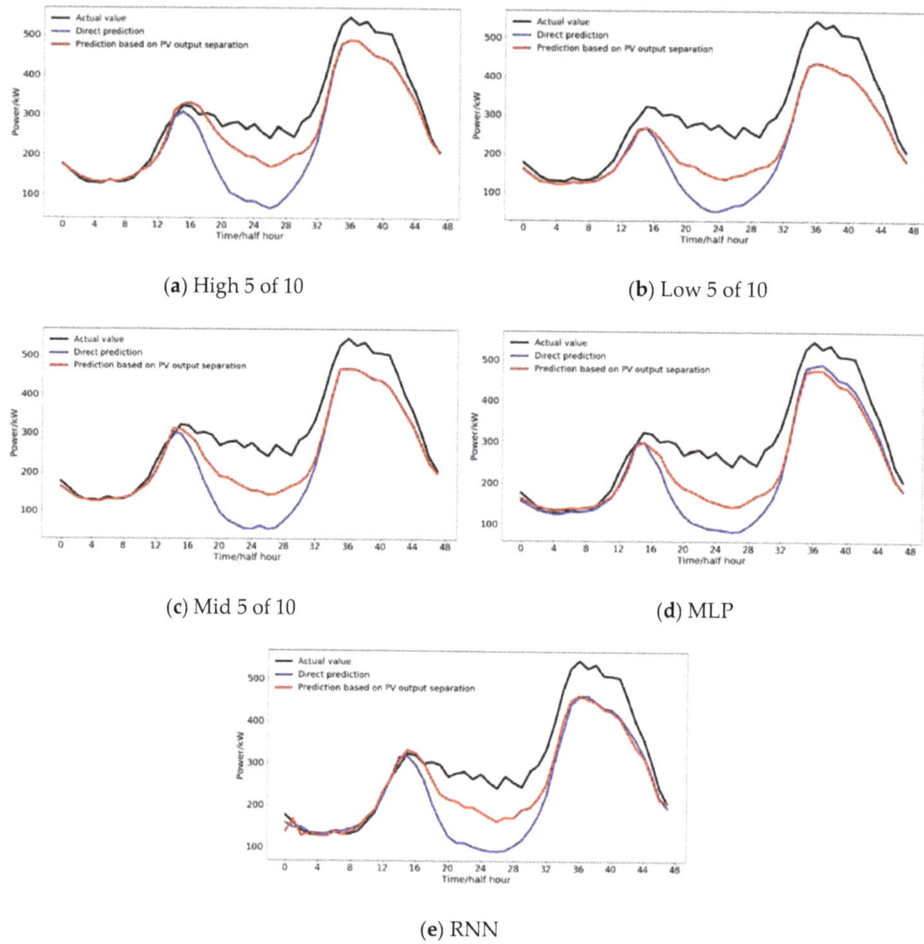

(**a**) High 5 of 10 (**b**) Low 5 of 10

(**c**) Mid 5 of 10 (**d**) MLP

(**e**) RNN

Figure 7. Comparison of direct prediction and customer baseline (CBL) prediction results based on PV output curve decomposition on a certain demand response (DR) day.

From the MAE and RER in Table 4, it can be seen that after applying the PV output power curve decomposition algorithm, the errors of the three CBL prediction methods are reduced and the proposed method has a better stability. This is because through the PV output power curve decomposition, the PV output part can get relatively accurate forecasting results. Regardless of whether the PV output power curve is decomposed or not, the bias in the table is negative, indicating that the predicted CBL is generally smaller than the actual value. One of the explanations is that the selected DR days have a larger load. Therefore, these five CBL prediction methods provided in the calculation example can only obtain smaller values than the actual CBL. At the same time, we can also see that because High 5 of 10 uses the larger data in the first ten days for CBL prediction, this method has the highest accuracy.

Table 4. Average evaluation results of CBL prediction for overall days under different methods.

CBL Prediction Methods	Direct Prediction			Prediction Based on PV Output Separation		
	MAE (kW)	Bias (kW)	RER	MAE (kW)	Bias (kW)	RER
High 5 of 10	73.46	−51.16	78.07	65.56	−45.44	79.34
Low 5 of 10	118.53	−115.00	79.28	102.05	−98.16	63.11
Mid 5 of 10	93.27	−82.33	77.19	80.36	−69.34	72.32
MLP	83.15	−74.76	76.44	74.49	−66.87	71.42
RNN	82.50	−71.69	72.42	69.15	−57.86	66.34

5. Conclusions

This paper proposes a decomposition method of the behind-the-meter PV output power curve. The advantage of this method is that it can estimate the PV output power curve data of aggregated customers when only the historical net load data, historical temperature data, and historical PV data of aggregated customers are known, without the need for knowing individual distributed PV output system equipment information. The framework firstly searches for similar time points in historical solar irradiation data, makes a difference to the corresponding net load, and obtains the temperature-based sensitivity of actual load according to the least squares method. Then it finds the date with similar electricity consumption behavior to match and uses the temperature-based actual load sensitivity to correct the net load of the matched day to the same temperature standard and makes a difference to offset the actual load part. Finally, MLP is used to fit the relationship between the delta solar irradiation and delta approximate PV output power to derive the PV output power curve estimation model. To illustrate the effectiveness of the proposed PV output decomposition framework, a total of 300 real customers' data sets containing PV output power from Ausgrid were used to simulate PV output decomposition experiments. At the same time, the results of the two types of prediction algorithms (average and regression) demonstrated that the CBL prediction based on PV output decomposition has better performance in MAE, bias, and RER.

In summary, this is a brand-new framework for BTM PV output estimation. Its most obvious characteristic advantage is that the algorithm does not depend on any data other than the historical net load data and weather data. In future work, the performance of the model will be analyzed more comprehensively, such as the impact of the resolution of the input data on the performance of the model, and the stability of the model under extreme weather conditions.

Author Contributions: Conceptualization, K.P., C.X. and C.S.L.; formal analysis, K.P., C.X. and C.S.L.; resources, C.S.L. and L.L.L.; data curation, K.P. and D.W.; writing—original draft preparation, K.P. and C.X.; writing—review and editing, C.S.L., L.L.L. and D.W. All authors have read and agreed to the published version of the manuscript.

Funding: This research received no external funding.

Acknowledgments: This work was supported by the Education Department of Guangdong Province: New and Integrated Energy System Theory and Technology Research Group [Project Number 2016KCXTD022]; the Guangdong Foshan Power Construction Corporation Group Co., Ltd.; Brunel University London BRIEF Funding, UK. The authors express their gratitude to the Editor for the kind invitation to contribute this research article, and to Joss Chen for her support in handling this submission.

Conflicts of Interest: The authors declare no conflict of interest.

Nomenclature

Abbreviations

BESS	Battery energy storage system	MLP	Multilayer perceptron
BL	Baseline	PV	Photovoltaic

BTM	Behind-the-meter	RAE	Relative absolute error
CBL	Customer baseline	ReLU	Rectified linear unit
DR	Demand response	RER	Relative error ratio
KNN	K nearest neighbor	RNN	Recurrent neural network
MAE	Mean absolute error	SCADA	Supervisory control and data acquisition
MAPE	Mean absolute percentage error		
Variables			
$\cdot(d)$	$\cdot$ of day d	n_s	Number of days in a season
$\cdot(s)$	$\cdot$ of season s	P_{ac}	Aggregated actual load
$\cdot(t)$	$\cdot$ of time t	$P_{ac,av}$	Average actual load
$\cdot(d,t)$	$\cdot$ of time t day d	$P_{ac,BL}$	CBL prediction of actual load
$\cdot(s,d,t)$	$\cdot$ of time t day d season s	P_{ac1}	Actual load during the time period with solar irradiation
ΔP^{*}_{ac}	Approximate actual load difference	P_{cor}	Correction of actual load difference
$\Delta P^{h*}_{ac,rad}$	Set of approximate delta actual load according to the date match I^{h}_{rad}	P_{est}	Estimated value of the PV output power
ΔP^{*}_{PV}	Approximate PV output power difference	P_{nl}	Aggregated net load
$\Delta P^{*}_{PV}{}'$	Correction of approximate PV output power difference	$P_{nl,BL}$	CBL prediction of net load
$\Delta P^{k*}_{PV,beh}$	Set of approximate delta PV output according to the date match I^{k}_{beh}	P_{nl1}	Aggregated net load data during the time period with solar irradiation
ΔR	Difference values between the solar irradiation of two days	P_{nl2}	Aggregated net load data during the time period without solar irradiation
ΔR^{k}_{beh}	Set of delta solar irradiation according to the date match I^{k}_{beh}	P_{PV}	Aggregated PV output power
ΔR^{h}_{rad}	Set of solar irradiance difference between two days with similar solar irradiance	$P_{PV,BL}$	Estimated PV output power for CBL prediction
ΔT_m	Temperature difference between two days	P_{real}	Real value of the PV output power
$\Delta T^{h}_{m,rad}$	Set of delta temperature according to the date match I^{h}_{rad}	r	Pearson's correlation coefficient
ρ	Euclidean distance	R	Solar irradiation
τ	Time period with solar irradiation	Rad_{amp}	Equivalent solar irradiation amplitude
$\overline{\tau}$	Time period without solar irradiation	$RAE_{PV,daily}$	Daily relative absolute error of PV output power estimation
d	day	s	Label of season
D	Set of day record	t	Time
I^{k}_{beh}	Recording set of the date of electricity consumption behavior similarity match	t_{rise}	Time of sunrise
I^{h}_{rad}	Recording set of the date of solar irradiation similarity match	t_{set}	Time of sunset
k	Number of the neighbors of KNN	T	Set of timestamp
$k_{temp,ac}$	Slope calculated by least squares method between delta temperature and delta actual load	$T_{m,av}$	Average temperature
MAE_{PV}	MAE of the PV output power estimation		

References

1. Wang, D.; Wu, R.; Li, X.; Lai, C.S.; Wu, X.; Wei, J.; Xu, Y.; Wu, W.; Lai, L.L. Two-stage optimal scheduling of air conditioning resources with high photovoltaic penetrations. *J. Clean. Prod.* **2019**, *241*, 118407. [CrossRef]
2. Lai, C.S.; McCulloch, M.D. Levelized cost of electricity for solar photovoltaic and electrical energy storage. *Appl. Energy* **2017**, *190*, 191–203. [CrossRef]
3. Zhao, Z.; Cheng, R.; Yan, B.; Zhang, J.; Zhang, Z.; Zhang, M.; Lai, L.L. A dynamic particles MPPT method for photovoltaic systems under partial shading conditions. *Energy Convers. Manag.* **2020**, *220*, 113070. [CrossRef]
4. Renewables 2020 Global Status Report–REN21. Available online: https://www.ren21.net/gsr-2020/chapters/chapter_01/chapter_01/ (accessed on 15 August 2020).
5. Lai, C.S.; Jia, Y.; Lai, L.L.; Xu, Z.; McCulloch, M.D.; Wong, K.P. A comprehensive review on large-scale photovoltaic system with applications of electrical energy storage. *Renew. Sust. Energ. Rev.* **2017**, *78*, 439–451. [CrossRef]

6. Hou, Q.; Du, E.; Zhang, N.; Kang, C. Impact of high renewable penetration on the power system operation mode: A data-driven approach. *IEEE Trans. Power Syst.* **2020**, *35*, 731–741. [CrossRef]
7. Xu, X.; Li, J.; Xu, Y.; Xu, Z.; Lai, C.S. A two-stage game-theoretic method for residential PV panels planning considering energy sharing mechanism. *IEEE Trans. Power Syst.* **2020**, *35*, 3562–3573. [CrossRef]
8. Xu, X.; Jia, Y.; Xu, Y.; Xu, Z.; Chai, S.; Lai, C.S. A multi-agent reinforcement learning based data-driven method for home energy management. *IEEE Trans Smart Grid.* **2020**, *11*, 3201–3211. [CrossRef]
9. Xu, X.; Li, J.; Xu, Z.; Zhao, J.; Lai, C.S. Enhancing photovoltaic hosting capacity-A stochastic approach to optimal planning of static var compensator devices in distribution networks. *Appl. Energy* **2019**, *238*, 952–962. [CrossRef]
10. Hippert, H.S.; Pedreira, C.E.; Souza, R.C. Neural networks for short-term load forecasting: A review and evaluation. *IEEE Trans. Power Syst.* **2001**, *16*, 44–55. [CrossRef]
11. Xu, F.Y.; Cun, X.; Yan, M.; Yuan, H.; Wang, Y.; Lai, L.L. Power market load forecasting on neural network with beneficial correlated regularization. *IEEE Trans. Industr. Inform.* **2018**, *14*, 5050–5059. [CrossRef]
12. Lai, C.S.; Mo, Z.; Wang, T.; Yuan, H.; Ng, W.W.Y.; Lai, L.L. Load forecasting based on deep neural network and historical data augmentation. *IET Gener. Transm. Distrib.* **2020**. [CrossRef]
13. Lai, C.S.; Locatelli, G.; Pimm, A.; Tao, Y.; Li, X.; Lai, L.L. A financial model for lithium-ion storage in a photovoltaic and biogas energy system. *Appl. Energy* **2019**, *251*, 113179. [CrossRef]
14. Lai, C.S.; McCulloch, M.D. Sizing of stand-alone solar PV and storage system with anaerobic digestion biogas power plants. *IEEE Trans. Ind. Electron.* **2017**, *64*, 2112–2121. [CrossRef]
15. Wu, X.; Lai, C.S.; Bai, C.C.; Lai, L.L.; Zhang, Q.; Liu, B. Optimal kernel ELM and variational mode decomposition for probabilistic PV power prediction. *Energies* **2020**, *13*, 3592. [CrossRef]
16. Lai, C.S.; Jia, Y.; McCulloch, M.D.; Xu, Z. Daily clearness index profiles cluster analysis for photovoltaic system. *IEEE Trans. Industr. Inform.* **2017**, *13*, 2322–2332. [CrossRef]
17. Siano, P. Demand response and smart grids—A survey. *Renew. Sust. Energ. Rev.* **2014**, *30*, 461–478. [CrossRef]
18. Shariatzadeh, F.; Mandal, P.; Srivastava, A.K. Demand response for sustainable energy systems: A review, application and implementation strategy. *Renew. Sust. Energ. Rev.* **2015**, *45*, 343–350. [CrossRef]
19. Xu, F.Y.; Zhang, T.; Lai, L.L.; Zhou, H. Shifting boundary for price-based residential demand response and applications. *Appl. Energy* **2015**, *146*, 353–370. [CrossRef]
20. Xu, F.; Huang, B.; Cun, X.; Wang, F.; Yuan, H.; Lai, L.L.; Vaccaro, A. Classifier economics of Semi-Intrusive Load Monitoring. *Int. J. Electr. Power Energy Syst.* **2018**, *103*, 224–232. [CrossRef]
21. Wang, D.; Qiu, J.; Reedman, L.; Meng, K.; Lai, L.L. Two stage energy management for networked microgrids with high renewable penetrations. *Appl. Energy* **2018**, *226*, 39–48. [CrossRef]
22. Zheng, J.; Lai, C.S.; Yuan, H.; Dong, Z.Y.; Meng, K.; Lai, L.L. Electricity plan recommender system with electrical instruction-based recovery. *Energy* **2020**, *203*, 117775. [CrossRef]
23. Li, K.; Wang, F.; Mi, Z.; Fotuhi-Firuzabad, M.; Duić, N.; Wang, T. Capacity and output power estimation approach of individual behind-the-meter distributed photovoltaic system for demand response baseline estimation. *Appl. Energy* **2019**, *253*, 113595. [CrossRef]
24. Wang, F.; Li, K.; Wang, X.; Jiang, L.; Ren, J.; Mi, Z.; Shafie-khah, M.; Catalão, J.P.S. A distributed PV system capacity estimation approach based on support vector machine with customer net load curve features. *Energies* **2018**, *11*, 1750. [CrossRef]
25. Jazaeri, J.; Alpcan, T.; Gordon, R.; Brandao, M.; Hoban, T.; Seeling, C. Baseline methodologies for small scale residential demand response. In Proceedings of the 2016 IEEE Innovative Smart Grid Technologies Asia (ISGT-Asia), Melbourne, Australia, 28 November–1 December 2016.
26. Zhang, X.; Grijalva, S. A data-driven approach for detection and estimation of residential PV installations. *IEEE Trans. Smart Grid.* **2016**, *7*, 2477–2485. [CrossRef]
27. Shaker, H.; Manfre, D.; Zareipour, H. Forecasting the aggregated output of a large fleet of small behind-the-meter solar photovoltaic sites. *Renew. Energy* **2020**, *147*, 1861–1869. [CrossRef]
28. Landelius, T.; Andersson, S.; Abrahamsson, R. Modelling and forecasting PV production in the absence of behind-the-meter measurements. *Prog. Photovolt. Res. Appl.* **2019**, *27*, 990–998. [CrossRef]
29. Tazvinga, H.; Xia, X.; Zhang, J. Minimum cost solution of photovoltaic–diesel–battery hybrid power systems for remote consumers. *Sol. Energy* **2013**, *96*, 292–299. [CrossRef]
30. Wu, K.; Zhou, H. A multi-agent-based energy-coordination control system for grid-connected large-scale wind-photovoltaic energy storage power-generation units. *Sol. Energy* **2014**, *107*, 245–259. [CrossRef]

31. Kabir, F.; Yu, N.; Yao, W.; Yang, R.; Zhang, Y. Estimation of behind-the-meter solar generation by integrating physical with statistical models. In Proceedings of the 2019 IEEE International Conference on Communications, Control, and Computing Technologies for Smart Grids (SmartGridComm), Beijing, China, 21–23 October 2019.
32. Shaker, H.; Zareipour, H.; Wood, D. A data-driven approach for estimating the power generation of invisible solar sites. *IEEE Trans. Smart Grid.* **2016**, *7*, 2466–2476. [CrossRef]
33. Cheung, C.M.; Zhong, W.; Xiong, C.; Srivastava, A.; Kannan, R.; Prasanna, V.K. Behind-the-meter solar generation disaggregation using consumer mixture models. In Proceedings of the 2018 IEEE International Conference on Communications, Control, and Computing Technologies for Smart Grids (SmartGridComm), Aalborg, Denmark, 29–31 October 2018.
34. Cheung, C.M.; Kuppannagari, S.R.; Kannan, R.; Prasanna, V.K. Disaggregation of behind-the-meter solar generation in presence of energy storage resources. In Proceedings of the 2020 IEEE Conference on Technologies for Sustainability (SusTech), Santa Ana, CA, USA, 23–25 April 2020.
35. Qiu, Y.; Kahn, M.E.; Xing, B. Quantifying the rebound effects of residential solar panel adoption. *J. Environ. Econ. Manag.* **2019**, *96*, 310–341. [CrossRef]
36. Pouraltafi-kheljan, S.; Göl, M. Power generation nowcasting of the behind-the-meter photovoltaic systems. *arXiv* **2020**, arXiv:2001.02157.
37. Chen, Y.; Xu, P.; Chu, Y.; Li, W.; Wu, Y.; Ni, L.; Bao, Y.; Wang, K. Short-term electrical load forecasting using the Support Vector Regression (SVR) model to calculate the demand response baseline for office buildings. *Appl. Energy* **2017**, *195*, 659–670. [CrossRef]
38. Wijaya, T.K.; Vasirani, M.; Aberer, K. When bias matters: An economic assessment of demand response baselines for residential customers. *IEEE Trans. Smart Grid.* **2014**, *5*, 1755–1763. [CrossRef]
39. Wang, F.; Li, K.; Liu, C.; Mi, Z.; Shafie-Khah, M.; Catalão, J.P.S. Synchronous pattern matching principle-based residential demand response baseline estimation: Mechanism analysis and approach description. *IEEE Trans. Smart Grid.* **2018**, *9*, 6972–6985. [CrossRef]
40. Park, S.; Ryu, S.; Choi, Y.; Kim, J.; Kim, H. Data-driven baseline estimation of residential buildings for demand response. *Energies* **2015**, *8*, 10239–10259. [CrossRef]
41. Haas, R.; Auer, H.; Biermayr, P. The impact of consumer behavior on residential energy demand for space heating. *Energy Build.* **1998**, *27*, 195–205. [CrossRef]
42. Hatton, L.; Charpentier, P.; Matzner-Lober, E. Statistical estimation of the residential baseline. *IEEE Trans. Power Syst.* **2016**, *31*, 1752–1759. [CrossRef]
43. Wu, X.; Wu, R.; Wang, D.; Wei, J.; Li, X.; Lai, L.L.; Lai, C.S. Coordinated air conditioning resources scheduling with high photovoltaic penetrations. In Proceedings of the 2018 International Conference on Power System Technology (POWERCON), Guangzhou, China, 6–8 November 2018.
44. Mu, Y.; Liu, X.; Wang, L. A Pearson's correlation coefficient based decision tree and its parallel implementation. *Inf. Sci.* **2018**, *435*, 40–58. [CrossRef]
45. Pinson, P.; Nielsen, H.A.; Madsen, H.; Nielsen, T.S. Local linear regression with adaptive orthogonal fitting for the wind power application. *Stat. Comput.* **2008**, *18*, 59–71. [CrossRef]
46. Keller, J.M.; Gray, M.R.; Givens, J.A. A fuzzy k-nearest neighbor algorithm. *IEEE Trans. Syst. Man Cybern. Syst.* **1985**, *15*, 580–585. [CrossRef]
47. Taud, H.; Mas, J.F. Multilayer Perceptron (MLP). In *Geomatic Approaches for Modelling Land Change Scenarios*; Springer: Berlin, Germany, 2018; pp. 451–455.
48. Kim, J.; Moon, J.; Hwang, E.; Kang, P. Recurrent inception convolution neural network for multi short-term load forecasting. *Energy Build.* **2019**, *194*, 328–341. [CrossRef]
49. Renewables.ninjia. Available online: https://www.renewables.ninja/ (accessed on 15 August 2020).

MDPI
St. Alban-Anlage 66
4052 Basel
Switzerland
Tel. +41 61 683 77 34
Fax +41 61 302 89 18
www.mdpi.com

Forecasting Editorial Office
E-mail: forecasting@mdpi.com
www.mdpi.com/journal/forecasting

www.ingramcontent.com/pod-product-compliance
Lightning Source LLC
LaVergne TN
LVHW070920170726
843515LV00005B/826

* 9 7 8 3 0 3 6 5 1 0 3 0 9 *